Building the Impact Economy

Maximilian Martin

Building the Impact Economy

Our Future, Yea or Nay

Springer

Maximilian Martin
Geneva
Switzerland

ISBN 978-3-319-25602-3 ISBN 978-3-319-25604-7 (eBook)
DOI 10.1007/978-3-319-25604-7

Library of Congress Control Number: 2015952536

Springer Cham Heidelberg New York Dordrecht London

Printed on acid-free paper

Springer International Publishing AG Switzerland is part of Springer Science+Business Media
(www.springer.com)

To my children Hara and Emanuela,
and our family
In the honor of our ancestors
To those we love and those who love us
And to those who follow
If we want them to have a history,
it is time to make it

Foreword by Bishop

In the aftermath of the financial crash of 2008, it was hard to believe that capital markets, or capitalism itself, could help make the world a better place. Yet in the years since then, a movement has been building, slowly at first but now gathering pace, founded on the conviction that it is not only possible for the capital markets to be harnessed to do good, but essential if the major challenges now facing the world are to be tackled successfully.

The phrase most often associated with this movement is "impact investment," meaning investment that simultaneously seeks to generate a financial return and deliver a demonstrable social and/or environmental benefit. In only a few years, impact investing has produced new sorts of companies, B Corps, some of them already worth over a billion dollars each. A new financial instrument, the pay for success "social impact bond," is catching on fast enough to encourage hopes that it could be the killer app that brings private capital usefully into areas now limited to nonprofit organizations and governments. There has even been a task force (of which I was the official report author) established by the governments of the G8 countries to look into how to catalyze a global impact investment market.

Moreover, impact investing is the cutting edge of a much broader set of trends, some of which stretch back decades, that emphasize responsible investment and invest where there is a good practice on matters of environmental and social and good corporate governance. As of summer 2015, some USD 57 trillion was managed by the investment firms that had signed up to the United Nations Principles of Responsible Investment, a pledge that at least theoretically binds signatories to take into account responsibly the social and environmental consequences of their investment choices.

Yet, as Max Martin argues persuasively in the chapters that follow, this movement will struggle to amount to much more than a promising start unless it becomes a good deal more ambitious about the speed and scale of its activities. A deliberate and entrepreneurial approach to harnessing the profit motive to do good needs to move rapidly from the fringes of the capital markets to their core.

I first met Max Martin over a decade ago, when he was at UBS, figuring out how the bank's philanthropically oriented clients could make a genuine difference through their giving. Conversations with him helped shape the ideas at the heart of "Philanthrocapitalism," the book I wrote with Michael Green that was published just days after the start of the financial crisis. Since then, we have spoken many times about what Max calls the "impact economy," as we have pursued similar questions about the role of profit in improving the world, and how philanthropists can work together with the for-profit capital markets.

Max has written here an important book, full of his characteristic insight and thoroughness. He rightly stresses the importance of raising the positive impact of corporate supply chains, especially in developing markets and of significantly raising the sums of capital going into renewable energy. After reading Max's words, you will be left in no doubt that impact investing has enormous promise and that significant additional resources need to flow into it if the world is to solve the huge problems now before us.

The G8 report on impact investing described the shift now under way as moving us from a two-dimensional approach based on risk and return to a three-dimensional way based on risk, return, and impact. Max has written an important contribution to our understanding of what this new 3D approach will require in practice.

New York
August 2015

Matthew Bishop

Matthew Bishop is a longtime writer for The Economist and the coauthor of several books with Michael Green, including "Philanthrocapitalism: How Giving Can Save the World."

Foreword by Schwartz

Recently, I returned from trips to two countries, Cuba, which I adore and visit regularly, and Bangladesh. They could hardly be more different in many ways. Yet, both Bangladesh and Cuba serve as a good example of the challenges and opportunities we face in this first quarter of the twenty-first century.

Cuba, a small country but the largest island in the Caribbean, is now re-establishing its relations with the USA. Direct flights between New York and Havana just resumed. People are beyond enthusiastic. They are also a bit wary of what the future might bring. Their culture and hospitality are legendary. The tourist industry and remittances have been of paramount importance since the collapse of the Soviet Union. Estimates say that the number of visitors will likely grow by another million over the coming year. Cuba has deep modernization and investment needs in agriculture, energy, and other sectors, which will be critical in order to be sustainably competitive in the global market. The country has also achieved remarkable universal health care and access to education and there is a strong desire to protect the humanitarian ideals that the revolution has struggled to achieve.

In Bangladesh I spent time with inspiring people working diligently for positive change in the country's booming garment industry. Without this industry, which accounts for four-fifths of export earnings, it would be difficult to see how Bangladesh can rapidly transform into the middle-income country that many of its 160 million citizens are anticipating. However, the tragic Rana Plaza and Tazreen tragedies among others also made it abundantly clear that life is still brutal and terribly precarious for far too many among us.

As an entrepreneur and someone trying to make a sustainable positive difference in the world, I strive to advance the causes of human dignity, social justice, and sustainability. As the former Executive Director of the Elie Wiesel Foundation for Humanity, I am keen to see us learn from past conflicts and build a world that is just and inclusive. I ask myself every day how I can contribute. I am deeply moved by the encyclical, Laudato si' in which Pope Francis reminds us:

> We have come to see ourselves as her [the Earth's] lords and masters, entitled to plunder her at will. The violence present in our hearts, wounded by sin, is also reflected in the symptoms of sickness evident in the soil, in the water, in the air and in all forms of life. This is why the earth herself, burdened and laid waste, is among the most abandoned and maltreated of our poor [1].

What can we really do then? Every purchase is a moral decision. Will that alone be enough? It is up to us to decide whether we should let our world crumble, or combine private initiative, our humanistic values, and modern financial engineering to avoid the Hobbesian world lingering around the corner.

If we are honest, our world is hanging in the balance. We urgently need new solutions. As a frequent speaker and author on effective global philanthropy and foundation asset management, I meet many high-impact entrepreneurs and social innovators in order to learn about their ideas and projects. This is why I am excited about *Building the Impact Economy*. Analytically stellar, the book gives us practical recommendations on how to get out of the current impasse by synthesizing responsibility and opportunity in an exciting new way.

In co-creating and building what emerged as the world's largest telecommunications bandwidth exchange, in my travels round the world, I learned firsthand about the power of technology as a force for good. Technology alone rarely does the trick though, because the people using it decide how much of a good or bad thing it can be. Maximilian Martin relates the power of technology to four themes. They are game-changing innovation in companies; a way forward for the global fashion industry so worker dignity is married with competitiveness; turning around the energy crisis; and restoring capital markets to being the funders of human progress and prosperity they once were—these are all essential pieces of the puzzle that is our future.

I hope that this book will shatter dogmas. With the current climate talks in Paris, its publication is especially timely—for it advances exciting insights of how to accelerate the energy transition as well. I am keen to see how these insights will play out. It is with great pleasure that I recommend this book as a most ambitious solution statement on how to best address these pressing challenges.

New York
August 2015

Daniel I. Schwartz

Reference

1. The Holy Father Francis (2015) Encyclical letter Laudato si' of the Holy Father Francis on care for our common home. http://w2.vatican.va/content/dam/francesco/pdf/encyclicals/documents/papa-francesco_20150524_enciclica-laudato-si_en.pdf. Accessed 15 July 2015

Daniel I. Schwartz is a co-creator of Arbinet and serves or has served on boards including Synergos, the Porticus Supervisory Board, the GAVI Campaign, the Arcus Foundation, the Rubin Museum of Art in New York, the Young Presidents' Organization (YPO), and the Chief Executives Organization.

Foreword by Leal

I am very happy to introduce this book. Drawing on the author's experiences over the last 15 years, the book builds a framework which shows how viable sustainable business approaches are, and demonstrates how much can be achieved if they are systematically implemented.

By zooming in on matters such as building businesses that add value to stakeholders and shareholders alike, and thus graduating from classical corporate social responsibility, the book shows how to move forward in a refreshingly clear way. It also analyzes how to modernize the world's #2 polluting industry—the global textile and garment industry—so it no longer destroys the environment and instead builds the competitive basis to pay workers sufficient income. The devotion of a substantial part of the book to energy and how to fast track the energy transition, a matter of central relevance to both industrialized and developing countries alike, reiterates the innovative nature of this publication, and is especially timely.

An issue seldom properly tackled is the influence of capital markets and the role that new investment models can play. Many of the current economic models are associated with insecurity, social exclusion, and sometimes lead to a decrease in levels of life quality and social well-being. Through his holistic thinking, combining attention to financial and non-financial returns, the author shows how investment and business development can promote environmentally sound, socially just and ethically acceptable growth. The current problems and crises reported by the media on a weekly basis show that this way of thinking is growing more urgent by the day. By combining "profits with impact" in ways that can become mainstream, the author shows how "impact investing" could develop a game-changing potential, and even lead to a fully fledged Impact Economy.

This book is important. It will make a tremendous difference in how we perceive current economic models, the role of investments and entrepreneurship. It will hopefully motivate us all to reflect on the many changes needed—and concretely outlined in this book—in order to change (and not perpetuate) the current status quo. The tools and methods are clearly explained. It is now up to us to run with them.

Hamburg
August 2015

Prof. Walter Leal

Prof. Walter Leal directs the International Climate Change Information Program (ICCIP) and has authored over 300 peer-reviewed publications. He served as a review editor of Working Group II of the Intergovernmental Panel on Climate Change (IPCC) Fifth Assessment Report.

Acknowledgments

Coming together is a beginning.
Keeping together is progress.
Working together is success.
Henry Ford [1]

Why "Building the Impact Economy"

This book provides concrete solutions how to get the world out of its current impasse.

In 2010, I initiated launch work that would result in the formation of the strategy and investment firm Impact Economy in early 2011. In October 2011, Swiss entrepreneur Christian Krüger backed the firm's vision as well—convinced by my pitch that we can and must take the market economy to a next level that fully reconciles economic, social, and environmental performance. Over the course of several conversations he shared how his personal interest in alternative energy had gotten a major impulse by direct exposure to the authors of *Limits to Growth* [2]. When the book was published in 1972 and the authors presented their work at the University of St. Gallen, where he was a student, the powerful framework showed for the first time the implications of unfettered use of resources in the complex, finite system that is our planet. As a general result, a large number of people suddenly realized that the world as we know it may in fact be headed for an "overshoot and collapse" scenario. This had important implications for action at the level of policy, business, and civil society.

Because my vision of Impact Economy seemed similarly bold, Christian Krüger challenged me to not only engage in a finite number of hopefully transformational strategy and investment projects, but to also describe my vision behind that work in a compelling narrative. The goal was to pull together my various bodies of work over the past fifteen years to come up with a comprehensive framework that had the potential to be of similar consequence.

Time to Be Ambitious

In the world of business, social and environmental considerations loom ever larger, yet a comprehensive framework for how to build sustainable new business approaches on a massive scale is still absent. It is moreover becoming clear that we are in a race against time, and that Corporate Social Responsibility (CSR), if anything, is an insufficient catalyst for major change. Global corporations, the business supertankers of the world economy, need a new and better approach to innovate. Many industries are becoming more and more global, and their supply chains longer, but beautiful products at the end of the pipe are routinely produced under horrible working conditions and result in endemic environmental damage at the front end, in the emerging markets where production takes place.

In the world of government, the Organization for Economic Cooperation and Development (OECD) debt crisis has increased the pressure for cheaper and more effective provision of public goods. Yet apart from isolated pilots such as the Social Impact Bond (SIB)—a structured financial product which systematically involves private investors in risk taking and capital mobilization for innovative performance-oriented public good provision—, an overall framework to rethink the welfare state in the twenty-first century in the context of sustainability and globalization is similarly missing. Such a framework would have to combine effectiveness with efficiency and engagement. Many governments are also looking for a meaningful role to enable a large-scale renewable energy transition. Without it, it is difficult to see how we could graduate from a cradle-to-grave to a cradle-to-cradle economy that uses, transforms, and recycles resources with the appropriate care. In a world heading for an expected nine billion inhabitants in 2050, closed-loop value creation will ultimately be the only way to reconcile human need and long-term sustainability.

Finally, in the world of investments, there is widespread concern that bankers have lost touch and no longer serve progress in the real economy. A new, to date still niche, movement for "impact investing" (i.e., investing with the intention to generate measurable social and environmental impacts as well as financial return) is gaining traction. If no remedial action is taken, it risks becoming another satellite activity focused on small-scale social entrepreneurs rather than transforming the mainstream of capital markets and investments.

Looking back at history, we can be confident that humanity has what it takes to sort out our problems. But we need to get the job done. While we personally act in biographical time, our actions influence historical time. Hopefully not unduly emboldened by encouraging conversations during which ideas always look easier to execute than is the case once one leaves the wine behind, I decided to take on the challenge, and to pull together a comprehensive intellectual and evidence-based framework. The goal was to synthesize these trends into a practical and motivating narrative that can help drive positive change on a large scale around the world, and that has the potential to provide a major platform to give visibility to new enterprise, investment and technology models, and innovations that are relevant for the transition.

While I recognize the power of the full spectrum of human emotions, I personally do not like to be scared. Going back to *Limits to Growth*, for me it followed readily that the point about the finite nature of our resources need not be made again. In fact, the world of scenario planning has developed in leaps and bounds over the past forty years. We have no scarcity of models to predict pretty much anything today, and to diagnose most problems. At the time, the authors wrote, "a decision to do nothing is a decision to increase the risk of collapse." [2, p. 188] This holds true in a system dynamics model with finite resources. I think that if we are asked to stop doing what we do, the way forward is to say what we should do instead. The good news is that, though scattered, there is a plethora of innovations out there, which can reconcile economic returns with social and environmental performance. We have not made comparable progress with respect to the questions though that I—as an entrepreneur who cares and has given a hand to several other innovations before—am particularly interested in: the *how-to* questions. If the problem is so bad and its urgency growing, what can we really do about it? How can we take our practice up not just one increment, but to a whole other level where success at the scale required all of a sudden becomes a serious possibility? How does this translate into products, services, and markets? The answers are by no means obvious.

To respond to what I believe are among the most relevant *how-to* questions today, I have chosen a few theaters and developed deep insights. The work on the book you are reading has taken almost four years. As any innovation process, it has been subject to a few false starts here and there. The resulting work is selective. It asks how we can step it up in four core domains: (1) Enabling large companies to innovate at the scale they potentially could. (2) Fixing long supply chains looking at a particularly problematic but alas somewhat representative example, the world of textile and garment manufacturing. (3) Asking how we can solve the energy crisis, because if we cannot get the mother of all markets right, there will be no closed-loop economy. And finally, (4) looking at how we can enlist capital markets to lend a helping hand to the transformation that we need to master if we want to reconcile long-term sustainability with private initiative and economic value creation. The epilogue offers thoughts on how human nature can give us a hand or obstruct the challenge we are facing—because this is about our future, yea or nay, so we better give it our best shot and succeed.

Sources of Support and Inspiration

The development funding by Krüger + Co. AG is gratefully acknowledged.

Moreover, the argument put forth in different parts of the book benefitted from deep subject matter insights gathered during projects for whose entrustment I am very grateful to the respective clients. Work on how large businesses can innovate benefitted from work supported by the Haniel Foundation. Work on how textile and garment manufacturing can be taken to the next level and how the energy transition can be fast-tracked benefitted from insights gathered during two-year-long projects mandated by undisclosed funders who provided the resources to enable me to ask hard empirical questions how to take our current practice up by more than just a notch. Finally, work on the role of impact investing in shaping the future of capital markets benefitted from my personal experience in setting up and running UBS Philanthropy Services—at the time the banking sector's first philanthropic services and impact investing offering for professional investors. Important insights also followed from research commissioned by the UK Cabinet Office/G8, who asked me precisely how they could stimulate market development in preparation of the first-ever G8 Social Investment summit presided by the UK Prime Minister as Chair of the G8 in 2013, and invited me to prepare a report providing a shared baseline for the summit's participants.

I am moreover deeply grateful to a number of individuals who have helped shape my thinking over the years. Here, I would like to especially thank Robert Glauber, my advisor at the Harvard Kennedy School who also offered me to serve as a Teaching Fellow in several of his courses, providing very valuable insights as to what to have on the radar when engaging in finance and capital markets work. They have been very helpful over the past decade and a half. Richard Zeckhauser has similarly informed my thinking about decision analysis and public policy. The Center for Public Leadership at Harvard University where I served as a Fellow more than a decade ago, and its founding director Ronald Heifetz, have similarly enabled lasting insights how we can not only diagnose a problem, but also organize convergent action to actually solve it. Moreover, I learnt a lot from Michael Jung at McKinsey & Company and Jürg Haller at UBS during the highly appreciated time worked for them at the two firms.

With respect to the manuscript of *Building the Impact Economy* proper, I am deeply grateful to all those individuals who were so kind to offer critical input on parts of the manuscript. Here listed in alphabetical order: Jeremy Birnbaum, Kim Brizzolara, Nick Busink, Martin Buttle, Rosendo Diaz, Andreas Ernst, Susy Guevara De La Torre, Walter Leal, Jean-Louis de Montesquiou, Philippe Sanlaville, Thomas Straub, Rainer Scheppelmann and Peter Waeber. Nick Busink, Andreas Ernst and Rainer Scheppelmann were so kind to provide their feedback on subsequent iterations of drafts, for which I am especially grateful.

Moreover, thanks for able research assistance are due to Egle Paulauskaite and Manijeh Torabbeigi at Impact Economy, who assisted for almost a year until the production of the first complete draft, and to Roxana Villalba for giving a hand in getting the manuscript ready for Springer, and the team at Springer for their valuable support in preparing the book for publication. I would also like to take this opportunity to thank our employees at Impact Economy, who—for the most part—worked hard to help make a difference on our projects, and to our Exergeia Fellows who helped us catalyze groundbreaking work on clean energy. As members of the

Generations X and Y, they will be among those reaping the benefits of success or suffering the consequences if we do not manage to reconcile economic, social, and environmental performance, and hit the proverbial iceberg instead, and were accordingly motivated.

References

1. Ford H. http://www.quoteground.com/authors/79-writers/211-henry-ford.html. Accessed 15 July 2015
2. Meadows DH et al (1972) The limits to growth. Universe Books, New York

Contents

Abbreviations

°C	Degree celsius
3D	Three dimensional
3G	Third-generation mobile technology
AAL	Ananta Apparels Ltd
Accord	Accord on Fire and Building Safety in Bangladesh
ADB	Asian Development Bank
AFWA	Asia Floor Wage Alliance
AIFMD	Alternative Investment Fund Managers Directive
Alliance	Alliance for Bangladesh Worker Safety
ARPA-E	Advanced Research Projects Agency-Energy
ASEAN	Association of Southeast Asian Nations
B2C	Business to consumer
BAU	Business as usual
BBC	British Broadcasting Corporation
BC	Before Christ
bcm	Billion cubic meters
BDT	Bangladeshi taka
BGL	Bangladesh
BGMEA	Bangladesh Garment Manufacturers and Exporters Association
BMZ	German Federal Ministry for Economic Cooperation and Development
Bn	Billion
BOI	Board of Investment Bangladesh
BoP	Base of the Pyramid (also referred to as the "bottom of the pyramid")
BP	British Petroleum
BRIC	Brazil, Russia, India, and China
BSC	Big Society Capital
BSCI	Business Social Compliance Initiative
BSR	Business for Social Responsibility
CAD	Canadian dollar
CAGR	Compound annual growth rate
CAP	Corrective Action Plan
CAR	Capital adequacy ratio
CCC	Clean Clothes Campaign

CdTe	Cadmium telluride
CEDIF	Community Economic Development Investment Fund
CEO	Chief executive officer
CIC	Community interest company
CIGS	Copper Indium Gallium (Di)Selenide [$Cu(In_xGa_{1-x})Se_2$]
CIS	Copper Indium Selenide
CIV	Corporate Impact Venturing
cm	Centimeter
CMT	Cut, Make, and Trim
CO	Cabinet Office
CO_2	Carbon dioxide
COGS	Cost of goods sold
CPV	Concentrated photovoltaics
CSP	Concentrating solar power
CSR	Corporate social responsibility
CVC	Corporate venture capital
DARPA	Defense Advanced Research Projects Agency
dB	Decibel
Defra	Department for Environment, Food and Rural Affairs
DFI	Development finance institution
DFID	Department for International Development
DIN	Deutsches Institut für Normung
DMF	N,N-dimethylformamide
DNA	Deoxyribonucleic acid
DNDi	Drugs for Neglected Disease initiative
DOD	US Department of Defense
DOE	US Department of Energy
e.g.	*exempli gratia*
EC	European Commission
ECHA	European Chemicals Agency
EDF	Electricité de France
EDI	Electric Drive Italia
EEA	European Environment Agency
EEG	Renewable energy law (in German, "Erneuerbare Energien Gesetz")
EIA	US Energy Information Administration
EIF	European Investment Fund
EIS	Enterprise Investment Scheme
EIT	European Institute of Innovation and Technology
EIU	Economist Intelligence Unit
EMIC	Ethiopia, Myanmar, Iran, and Colombia
EPR	European Pressurised Reactor
EPZ	Export processing zone
ESG	Environmental, social, and governance
etc.	*et cetera*
ETI	Ethical Trading Initiative

EU	European Union
EUR	Euro
EV	Electric vehicle
FCA	Financial Conduct Authority
FCEV	Fuel cell electric vehicle
FCPR	Fonds commun de placement à risques (French venture capital fund)
FDI	Foreign direct investment
FHS	Fondazione Housing Sociale
FIA	Fondo Investimenti per l'Abitare (Italian social housing fund)
FLA	Fair Labor Association
FMCG	Fast moving consumer goods
FMO	Netherlands Development Finance Company
FOA	Funding Opportunity Announcement
FTSE100	Financial Times Stock Exchange 100 index
FWF	Fair Wear Foundation
G8	Canada, France, Germany, Italy, Japan, Russia, United Kingdom, United States of America
GBP	Pound sterling
GDP	Gross domestic product
GE	General Electric Company
GHG	Greenhouse gas
GIIN	Global Impact Investing Network
GIIRS	Global Impact Investment Rating System
GIZ	Deutsche Gesellschaft für Internationale Zusammenarbeit
GMAC	Garment Manufacturers Association in Cambodia
GOTS	Global Organic Textile Standard
GPS	Global Positioning System
GRI	Global Reporting Initiative
GSP	Generalized System of Preferences
GTP	Growth and Transformation Plan
GW	Gigawatt
HAW	High-altitude wind
HM	Her Majesty
i.e.	*id est*
Ibid.	*ibidem*
ICCIP	International Climate Change Information Program
ICRF	Investment and Contract Readiness Fund
IDB	Inter-American Development Bank
IEA	International Energy Agency
IFC	International Finance Corporation
IFOAM	International Foundation for Organic Agriculture Movements
IIC	Inter-American Investment Corporation
ILO	International Labour Organization
ILRF	International Labor Rights Forum
IMF	International Monetary Fund

IoT	Internet of Things
IPE	Institute of Public & Environmental Affairs
IPO	Initial public offering
IRENA	International Renewable Energy Agency
IRIS	Impact Reporting and Investment Standards
IRR	Internal rate of return
ISO	International Organization for Standardization
IT	Information Technology
ITER	International Thermonuclear Experimental Reactor
ITU	International Telecommunications Union
kg	Kilogram
KIC	Knowledge and Innovation Community
km	Kilometer
KPI	Key performance indicator
KSU	Kite Steering Unit
kWh	Kilowatt-hour
L3C	Low-profit limited liability company
LCICG	Low Carbon Innovation Co-ordination Group
LCOE	Levelized cost of energy
LED	Light emitting diode
LEED	Leadership in Energy and Environmental Design
LENR	Low-energy nuclear reactions
LOHAS	Lifestyles of Health and Sustainability
LP	Limited partner
MA	Master of Arts
MDGs	Millennium Development Goals
MEP	Ministry of Environmental Protection of the People's Republic of China
MGMA	Myanmar Garment Manufacturers Association
MIC	Myanmar Investment Commission
MIF	Multilateral Investment Fund
MiFID	Markets in Financial Instruments Directive 2004/39EC
MIT	Massachusetts Institute of Technology
MJ	Megajoule
MRI	Mission-related investment
MTF	Multilateral trading facility
MTN	Mobile Telephone Networks
MTOE	Million tons of oil equivalent
MW	Megawatt
MWh	Megawatt-hour
n.d.	Not dated
N-11	Next-11
NASA	National Aeronautics and Space Administration
NBE	National Bank of Ethiopia
NECT	Nifurtimox-Eflornithine Combination Therapy

NGO	Non-governmental organization
NIF	New Initiatives Fund
NMI	Natural Marketing Institute
NPE	Nonylphenol ethoxylate
NREL	National Renewable Energy Laboratory
NTPA	National Tripartite Plan of Action on Fire Safety and Structural Integrity
OBM	Original brand name manufacturing
ODA	Official development assistance
ODM	Original design manufacturing
OECD	Organisation for Economic Co-operation and Development
OEM	Original equipment manufacturing
OHS	Occupational health and safety
OMJ	Opportunities for the Majority (initiative)
p	Page
P2P	Peer-to-peer
PDP	Product development partnership
PPR	Pinault-Printemps-La Redoute
PRI	Program-related investment
PU	Polyurethane
PwC	PricewaterhouseCoopers
R&D	Research and development
RAGS	Responsible and Accountable Garment Sector
REACH	Registration, Evaluation, Authorisation and Restriction of Chemicals
RFID	Radio-frequency identification
RMG	Ready-made garment
SAC	Sustainable Apparel Coalition
SAI	Social Accountability International
SBIC	Small Business Investment Company
SBIR	Small Business Innovation Research
SCF	Structured and Corporate Finance
SCORE	Sustaining Competitive and Responsible Enterprises
SEC	US Securities and Exchange Commission
SEF	Structural, electric, and fire safety
SGS	Société Générale de Surveillance
SIB	Social impact bond
SITC	Standard International Trade Classification
SME	Small- and medium-sized enterprises
SRI	Socially responsible investment
SVHC	Substance of very high concern
SVP	ShriVallabh Pittie Group
ToP	Top of the pyramid
TPES	Total primary energy supply
TPP	Trans-Pacific Partnership
TWh	Terawatt-hours

UK	United Kingdom
UN	United Nations
UNEP	United Nations Environment Program
UNICEF	United Nations Children's Fund
URL	Uniform Resource Locator
US	United States
USA	United States of America
USD	United States Dollar
V2G	Vehicle-to-grid
VC	Venture capital
Vol	Volume
vs.	Versus
W	Watt
WB	World Bank
WBCSD	World Business Council for Sustainable Development
Wh	Watt-hour
WHO	World Health Organization
WRAP	Waste and Resources Action Program
WTO	World Trade Organization
YPO	Young Presidents' Organization
ZDHC	Zero discharge of hazardous chemicals

Testimonials

"Maximilian Martin is able to capture the urgency as well as the systemic nature of change in his brilliantly practical and solutions oriented book 'Building the Impact Economy.' This urgency for meaningful transformation is shared by thousands of impact investors who see their impact portfolios not as an exercise to make more money, or even as proof-points for meaningful impact, but as an expression of who they really are and the change they want to see. In this book, Max has put together a wonderfully illuminating and practical roadmap for the world's impact investors and their investees."

—Charly Kleissner, Ph.D., Co-Founder KL Felicitas Foundation, Toniic, 100 % IMPACT Network, Social-Impact International

"This book provides a highly ambitious, actionable, and readable framework to get us from mere diagnosis of the world's problems to all-encompassing, real steps forward, including with the energy transition. A major contribution to aligning entrepreneurial value creation with the ideal of the honorable merchant in our world today, and most valuable guidance for professionals and talent of any age who want to make a difference in times of rapid change."

—Rupert Antes, Executive Director, Haniel Foundation

"As an entrepreneur, I believe in finding innovation. As a philanthropist, in innovation to change the reality for those who are less privileged. Maximilian Martin provides a most stimulating synthesis how new value propositions in business and investment can tick both boxes, and how to concretely come up with them. Required reading for my fellow entrepreneurs."

—Yann Borgstedt, entrepreneur and Founder, Womanity Foundation

"Building the Impact Economy is striking both for its originality and for its contribution to the policy debate on new approaches to old problems. Connecting sustainable business innovation, global supply chains, renewable energy, and capital markets, it sheds a new light on the massive potential of social impact investing. Martin also shows the crucial role of smart government in enabling innovation ecosystems. Policymakers and investors will enjoy the many actionable insights and case studies."

—Kieron Boyle, Head of Social Investment, Government Innovation Group, UK Cabinet Office

"An inspiring and illuminating roadmap for the way we run, and invest in, the market economy. Martin comes up with new win-win scenarios that will help sort out the great environmental and social problems that are facing us. He suggests doing so by leveraging private initiative, science and capital markets. The women who work in global supply chains and produce the clothes we wear are among those who stand to benefit the most. The book's many practical recommendations being implemented with the ambition level now required—that's a story I would like to tell."

—Kim Brizzolara, feature film and documentary producer and private investor, Executive Vice Chair of the Hampton's International Film Festival

"The next decade is likely to be a period of profound change for organizations working in the social sector as private capital, innovation and entrepreneurship are increasingly channeled into delivering public goods and solving problems. Maximilian Martin has been working at the cutting edge of impact investing for many years and draws extensively and insightfully on this experience in his new book. A very important read for current and aspiring leaders who are seeking to drive social impact, as this nascent financing revolution gathers pace and changes the rules of the game."

—Gib Bulloch, Founder, Accenture Development Partnerships

"Maximilian Martin has written the best book that I have seen on enlisting sustainable investing to solve today's urgent problems in energy, upgrading manufacturing in emerging markets, and business innovation. His writing is lively, full of relevant case studies, and rich in insights. A must read for novice and seasoned sustainable investors alike."

—Andreas Ernst, Head of Impact Investing, UBS

"Capital markets have a unique power to direct resources to the new win-win of profit and impact. Based on a very thorough analysis of the opportunity, Martin develops a most compelling roadmap how to make sustainable business and investing the main show in town. Exactly what's needed if we want to get out of the current impasse, and required reading for anyone who cares and invests money."

—Alois Flatz, Co-Founder Dow Jones Sustainability Index, and Venture Partner, Zouk Capital LLP

"A must-read for members of the online tribe who are asking themselves how our digital prowess can help us make real progress where it hurts: global supply chains, the energy transition, big business and capital markets. For business and investing, the book is just as disruptive as the Internet of Things was for the world of digital. Turning around the ship has just become so much more compelling."

—Andreas Gebhard, Co-Founder and CEO of re:publica and Co-Founder of newthinking, Berlin

"This book is important. It will make a tremendous difference in how we perceive current economic models, the role of investments and entrepreneurship. It will hopefully motivate us all to reflect on the many changes needed—and concretely outlined in this book—in order to change (and not perpetuate) the current status quo. The tools and methods are clearly explained. It is now up to us to run with them."

—Professor Walter Leal, Director International Climate Change Information Program (ICCIP), and review editor of Working Group II of the IPCC Fifth Assessment Report

"A vademecum in today's world of business and finance, where we know much is at stake but are often unsure how to deal with the root causes. Martin persuasively argues that the new win-win must use technology and innovative financing to achieve sustainability and competitiveness goals at the same time. Essential reading for all those of us who set up the structures of corporate entities and foundations that condition practice long after the founders have left the bridge."

—Benoît Merkt, Partner, Lenz and Staehelin

"I hope that this book will shatter dogmas. With the current climate talks in Paris, its publication is especially timely—for it advances exciting insights how to accelerate the energy transition as well. It is with great pleasure that I recommend this book as a most ambitious solution statement on how to best address these pressing challenges."

—Daniel Schwartz, co-founder of telecommunications bandwidth exchange Arbinet

"As a serial entrepreneur, I built many businesses in energy and other fields—always working hard to get both profits and values right. Max's inspiring book shows how to systematically amplify individual genius and commitment to engineer new business and investment propositions in the key theaters where we need to move beyond talk about solving the world's problems. Electrifying reading that answers the entrepreneur's how-to questions, and essential for anyone who wants emerging markets to truly thrive."

—Rory Stear, serial entrepreneur, Executive Chairman, Flambard Capital, United Kingdom, and Chairman of the Board, Business School, Nelson Mandela University

"A highly innovative and promising approach, which takes into account the changing division of labor between business, civil society and government in addressing our world's most urgent challenges, and proposes to step up solution formulation using modern finance. Managers and investment professionals should read this book as a practical roadmap for innovation that matters."

—Thomas Straub, Professor of International Management and Director of the Executive Education, Faculty GSEM, University of Geneva

"Dr. Maximilian Martin has created a work that not only comes quickly to the point but also shows decision-makers in the area of sustainability with supply chain responsibility a vision that is both feasible and has implementation potential. 'Impact Economy' is exactly the key with political practicability not only to discuss changes but to also realize them. I recommend this work to all 'movers and shakers' in the industry to start the long overdue adjustments regarding the gap between humanity's demands on resources and the planet's biocapacity."

—Peter Waeber, Founder and CEO bluesign technologies, Switzerland

List of Spotlights

List of Figures

List of Tables

Prologue: What's at Stake?

If you want to go fast, go alone.
If you want to go far, go together.
African Proverb [1]

A New World Order—So What?

You have started to read this book because you are interested in the new order of things that is now emerging—is the post-World War II market economy viable in our crowded, agitated, and resource-constrained twenty-first century? What's really at stake? What needs to change so our world can be somewhat humane and remain viable? How can we avoid first overshooting at some point, to only then fall back to a lower level equilibrium, as many scenario planners have cautioned ever since the publication of *Limits to Growth* more than forty year ago? [2].

This book proposes the solutions to get us out of our current impasse. The contours of a world order successfully combining private initiative, markets, and sustainability are gradually becoming clearer. But if the innovations that are currently being rolled out do not gain scale fast enough, and if we do not build a pipeline of powerful solutions for the most pressing problems of our time—which we are not yet able to resolve in spite of our ability to perform a soft landing on a comet and return data from its surface after a ten-year journey through space (i.e., the Rosetta Mission) , a fairly ugly world might instead materialize just as well.

In other words: It is about our future, yea or nay. This book argues that our best shot to end up in the "yea" camp is to build an Impact Economy. It asks the *how-to* questions that will permit us to get there. The macro-view is pretty clear. Even if current zones of prosperity could be defended with ever more sophisticated military technology, the environmental limits of our planet will eventually catch up with us. The current rise of extremist social movements and periodical sudden eruption of social conflict in East and West give us a flavor of what could happen on a much larger scale if we fail to build inclusive, competitive and sustainable economies and societies around the world.

When world orders change, chaos is the most frequent consequence. Historical precedents abound. Part of my ancestry is Frankish, and part Hanseatic; there are conclusions to be drawn from the intertwined history of the Germanic tribes and the Roman Empire. Take Clovis I, whose father was a sound federate of the Romans against the Visigoths in Gaul. He expected the Romans to upgrade his role in recognition of the Franks' services as an ally. This never happened; rather, the North of the former province of Gaul was cut off from Italy in the Migration Period. Clovis I turned and defeated the Roman magister militum in the Battle of Soissons in 486. Had the Romans found a way to integrate the barbarians who were putting pressure on the borders of their depopulating empire, the history of Europe could have played out differently.

Opportunity and a shared dream are the only way to provide an environment where a critical mass of citizens has a stake in the system and actively contributes to upholding it. Personally, I do not consider a world desirable where crisis management of expanding social upheaval, event-driven management of environmental disaster, and defending zones of prosperity by deploying advanced military robots combatting hordes of extremists to avoid casualties, become the new normal. Nor do I believe that such a strategy would actually be viable in the end. Just consider how the current migration from Africa across the Mediterranean is putting pressure on Europe. The implications are scary for everyone: for Western liberal democracy; for the societies of origin that lose their most enterprising people; and for the refugees who often suffer personal tragedies when they embark on small boats to cross the seas, responding to desperation with daring.

To master the multitude of challenges we face, what we really need are fresh ideas. And we need equally clever ways to go about implementing them. Real solutions will require new frameworks: "Problems cannot be solved by the level of awareness that created them," as Albert Einstein famously argued. Crosscutting principles integral to any viable solution are that it needs to (1) transcend short-termism, (2) shift the focus from remedy to solution, (3) grasp the systemic interconnectedness of problems and remedies, and (4) activate levers of improvement where a limited input unlocks disproportionate benefit, making the system feed on itself.

This book looks at several core arenas of action to have on our radar if we are serious about building an alternative: a world where zones of prosperity gradually expand and a sustainable economy goes mainstream. It's fairly straightforward: (1) if economic value creation increasingly pays attention to the social and environmental impacts it creates, (2) if the social sector improves its productivity and effective use of resources, and (3) if government systematically uses its bully pulpit to underwrite future-oriented policies without crowding out private initiative, then we are talking. I have written this book as a primer for the people who need to make this happen: every one of us, the "Great Generation" that needs to help set us on the right course as the first quarter of the twenty-first-century unfolds. We could also call it "Generation Impact." The good news is that private enterprise will be a key success factor: Building the Impact Economy will also result in numerous profitable activities and investments.

Many of the building blocks that we need to draw on to achieve this outcome already exist. However, for now they are islands, not continents. Their proponents are sometimes so busy fighting with colleagues over details that they lose sight of the big picture. For example, one idea that started to gain traction in the early

2000s is to leverage capital markets for positive change. The core idea is this: systematically assess the nonfinancial by-products of investment activities and integrate them into your investment decisions. In 2010, USD 600 trillion in total financial assets globally represented claims against a nominal global Gross Domestic Product (GDP) of roughly USD 63 trillion [3]. Surely, one should be able to fast track positive change if one could tap into that capital pool. Referred to as "impact investments"—investments made with the intention to generate measurable social and environmental impact alongside a financial return—, this is one of the topics covered in Part IV [4]. Even if they reach the growth target of one trillion dollars estimated as doable by specialists, this would still only correspond to 0.1 % of all estimated financial assets by 2020. Under a business as usual scenario, it is hard to imagine how such a small tail could wag the proverbial dog of global finance gone haywire and reconnect it to providing the capital needed to build strong sustainable and profitable economies. This means we need to go mainstream faster.

The general premise of this book is that it is time to aim higher. To have a reasonable chance of success, we need to develop both a sophisticated understanding of the different levers and possibilities in play, and how they connect to the overall future a twenty-first-century humanist would like to see materialize. The questions are (1) how to make holistic innovations combining private initiatives and markets with the greater good happen in practice, (2) where the frontiers are, and (3) which key opportunities we must act on now. The "impact revolution" is well under way. The goal now is to make it the main show in town. We will not look at everything, but at the key theaters.

Some of what you will read may sound futuristic: the use of Bitcoins in global finance, solar panels with an up to 70 % efficiency, or wet processes in textiles that save up to 95 % of water. To some, the argument may moreover be controversial and even invite criticism by those who have a stake in "business as usual" (BAU)—perhaps because they will not be around when BAU hits the wall. Or because they have not taken the time to work through its implications.

I have nevertheless written this book based on the assumption that stirring such criticism can be a sign of success, not failure. The goal was to trigger engagement with issues we need to deal with [5]. This requires speaking truth to power. Working on issues of innovation and capital markets over the past fifteen years has made me weary of the type of language you will find in the disclaimers of investment proposals. However, I can assure the reader that all information in this book has been obtained in good faith. It is validated by practical work and insights into different theaters covered well exceeding 10,000 h.

Before you read any further and whether you consider yourself an innovator or not, please do bear in mind that most people like innovations and innovators only after the fact. They fight against their ideas and solutions until they have gained widespread acceptance. Insights from an old acquaintance who took the world of politics to another level in his day remain as topical now as they were then. 500 years ago, Niccolò Machiavelli offered relevant pointers as to what one is up against if one wants to make changes to some current system [6]:

> It ought to be remembered that there is nothing more difficult to take in hand, more perilous to conduct, or more uncertain in its success, than to take the lead in the introduction of a new order of things, because the innovator has for enemies all those who have done well under the old conditions, and lukewarm defenders in those who may do well under the new. This coolness arises partly from fear of the opponents, who have the laws on their side, and partly from the incredulity of men, who do not readily believe in new things until they have had a long experience of them.

Business Enabling Sustainable Innovation (and Profits)?

In this prologue chapter, let us now quickly review the main topics in the plot, which you will encounter unfolding in this book. Unsurprisingly, one important player in building sustainable economies and societies are companies, and especially the large ones. These "supertankers" of the world economy are large and have clout. They turn out products and services on an impressive scale and generate significant amounts of wealth and tax revenue. But they also suffer from the inertia of any object with great mass, as we remember from physics in high school.

Corporations can make major contributions. Nevertheless, to do this on a scale that is interesting, they first need to free themselves from the limiting dichotomy between corporate responsibility and corporate opportunity, as practiced under the heading of "corporate social responsibility" (CSR). CSR can be an exercise where more brainpower and resources are spent on communicating how virtuous a company is than on the underlying community projects themselves. Economically absurd resource allocation patterns such as having oil engineers clean up beaches or managers painting schools can be common practice. This rather than asking what kind of relevant skill set employees can bring to bear on a social or environmental issue so they achieve more than feeling good about themselves and producing fodder for the communications department.

Part I has some good news however: More and more consumers want companies to address social and environmental problems through their product offering (rather than see them engage mainly in at best non-scalable and at worst token social projects). This is one of the reasons why the days of traditional CSR are gradually coming to an end. We can start to make sense of a new paradigm emerging in its place that is responding to structural changes in the operating environment of business. Deploying best available technologies and business model innovation to render global supply chains both more competitive *and* more humane is more effective than most conceivable CSR programs.

There are several megatrends that significantly shape value creation, including the five trillion dollar markets at the Base of the Pyramid (BoP), the 500+ billion dollar market for Lifestyles of Health and Sustainability (LOHAS) consumption, the multitrillion dollar markets resulting from green growth and the emerging circular economy, as well as the modernization of the welfare state [7]. To cite just two examples, we have four billion poor potential consumers around the world (with aggregate demand for one billion toilets alone), as well as a 1.7 billion strong emerging middle class. We better keep them on the radar. They also want to enjoy the mobility, housing, health, and education, all of which became widespread in the West after World War II. As people in emerging markets redefine what business relevance actually means, the consequences could be virtuous or disastrous from a social and environmental perspective [8].

Companies from developed countries will have to learn how to tap into the changing drivers of global growth. When McKinsey & Company surveyed 100 leading companies from developed countries in 2012, it found that 83 % of revenues came from developed markets despite the fact that the share of global GDP of emerging markets had nearly doubled from 19 to 36 % in just two decades [9, 10]. Part I will look at the shifting balance, moving on from an exclusive focus on responsibility to one that assesses joint opportunities for financial and social returns. By updating patterns of value creation in new and existing multitrillion dollar markets, this approach has significant potential in terms of scale and impact. Provided its endeavor to achieve actual financial returns, greater effectiveness, and more leverage potential for both business and society, is successful, corporations could make a key contribution to building an inclusive market economy.

Fashion: Breaking Out of Dress to Kill

When we talk about identifying and rolling out the innovations needed to build an inclusive, viable global economy and society, the rubber hits the road where real people doing mainstream jobs and products we use every day are concerned. Part II looks in detail at an industry that perhaps best captures the contrast between the beauty of its products and abysmal working conditions: global fashion.

Today, the global textile and garment industry is at a crossroads. It is a three-trillion-dollar industry that encompasses the manufacturing and selling of textiles and garments. It has long been considered a source of economic progress around the world [11]. Historically, the industry has served as a catalyst for national development and industrialization. The flipside of this growth and the accelerating production of fashion has been a broadening and deepening track record of poor working conditions and heavy pollution. The harsh realities of the workplace still fail to match the beauty of the products they manufacture; the glamorous world of fashion and the world of making it can be brutally contrasting.

The collapse of the Rana Plaza factory in April 2013 in Dhaka, Bangladesh (BGL), killing over 1100 garment workers, jolted to life widespread and increasingly prolonged scrutiny of the industry. This incident has brought longstanding questions to the forefront over how to bridge the gap between economic viability and social and environmental performance. Part II sorts through the minefield of perspectives, expectations, and challenges that have only grown more complicated since Rana Plaza. It provides an evidence-based assessment of the prospect of sustainable value chains in the textile and garment industry. We ask the ultimate question on how the industry can substantially raise its social and environmental performance, while staying competitive.

Once again, there is good news. To tackle its working conditions with a new level of ambition, and to redesign production, improve infrastructure, and provide workers and management the training they need to boost their productivity—so that value added can further rise—is not a utopian ambition. Linking greater competitiveness with improved social and environmental performance in the clothing industry is possible. In today's global apparel industry, surprisingly many sources of potential economic, social, and environmental productivity gains remain untapped. Concretely, we can save up to 20 % of chemical inputs, up to 40 % of energy, and up to 50 % of water in the manufacturing process [12]. Addressing factory working conditions can yield competitiveness dividends and social impact —a win–win situation for everyone. This could mean better workplace environments for women who make up over 80 % of the workforce—and this makes the fashion industry a relevant placeholder for labor-intensive industries with global supply chains in general. If we cannot take them to a new level, it is difficult to see how a sustainable world could emerge in the near term.

Fixing Energy, the Mother of all Markets

By 2019, the Prime Minister of India, Narendra Modi, aims to bring solar power to the 400 million Indians currently lacking access, so that every home will have access to electricity [13]. A laudable and highly ambitious goal, it also tallies well with a global community engaged in global climate negotiations, potentially setting important parameters to limit carbon emissions.

Throughout history, access to energy has ranked at the same level of importance as basic human priorities such as food, housing, health, or education. Energy is foundational to the ability to provide all of the former. Access to energy will remain central to the definition of geopolitical fault lines, as haggling over pricing of natural gas exports from Russia or wars over oil fields in the Middle East remind us nearly every day. A successful energy transition that occurs fast enough to reduce pressure on our environment is not a foregone conclusion however.

In a perfect world, renewables would turn out electricity at a few cents a kilowatt-hour (kWh), solar panels would capture energy at night, and storage solutions would compete with gasoline in terms of energy density and ability to release energy. If the energy transition is to succeed in India and elsewhere, renewable energy will have to become fully price competitive and substitutable with fossil fuels much faster. The way forward is to deploy subsidies to finance innovation and to reach critical mass only, rather than providing subsidies when operating at scale.

Storage remains a sticky point though, in terms of battery performance as well as the ability to roll out the state-of-the-art products. Immediately following their launch on April 30, 2015, Tesla—the premium electrical car manufacturer—, received 38,000 reservations for two new high-performing batteries called Powerwall (for residential use), and 2500 for PowerPack (for industrial use). If all of Tesla's Gigafactory production were devoted to just producing stationary batteries, then Tesla could fulfill current reservations by mid-2016. A delivery constraint for Tesla, storage becomes a key constraint for the energy transition, seen that renewable sources produce energy only intermittently.

Energy storage plays a key role to integrate more renewables into the grid, for continuous and flexible supply of energy and for reducing energy costs by peak shaving. A breakthrough in batteries would be very helpful for India's rural solar revolution. In urban areas, it would be a prime enabler of e-mobility. The problem is however: Today's batteries are not yet up to the task. Moreover, wouldn't it be great if the best ones came to market faster?

In Part III, we will look at what needs to happen to REALLY fast track the energy transition. For example, the major challenges today's batteries face are cost, lifespan, safety, as well as performance over a wide temperature range (−30 to 52 ° C). For example, the chemical energy stored in 1 kg of coal is 24 MJ/kg. Combusting this coal in a thermal power plant with an efficiency of 40 % would correspond to a useful energy of 9.6 MJ/kg of coal. By contrast, the energy density in the Tesla S model, perhaps the most advanced commercially available lithium-ion technology, corresponds to 233 Wh/kg (0.84 MJ/kg). Batteries currently used in hybrid and electric cars average only 150 Wh/kg, compared with gasoline's 12,000 Wh/kg (44 MJ/kg).

That is far from good enough. Progress in storage compares very unfavorably with, say, advances in semiconducting: Rather than progress on a logarithmic scale in computing power, we are currently looking at progress on a linear scale in storage. Based on a nine-month comprehensive global scouting effort for the most promising clean energy solutions and technologies around the world (and follow-on work), I led via Impact Economy's Exergeia Project, we identified and screened over 9000 start-ups and projects. The goal was to assess their potential to compete on a non-subsidy basis with fossil fuels. Drawing on the insights gathered, Part III identifies the leading types of opportunities to shift out the possibility frontier, and asks what investors, inventors, and the public sector can do to move forward.

Capital Markets: Graduating from the Casino

The financial crisis called into question business as usual and whether global capital markets effectively fulfilled their role of pricing risk and allocating capital. The notion of safe assets is now gone and financial markets are often seen as "casinos" instead. In Japan alone—the world's third-largest economy—, USD 8.5 trillion sits in commercial banks as cash deposits [13]. There will be an estimated USD 900 trillion in financial assets by 2020, representing claims against about USD 90 trillion/year of global GDP [13]. The figures in the world of global finance are so large that even transformational initiatives in relatively large developing countries, such as the total estimated USD 75–79 billion cost of Ethiopia's Growth and Transformation Plan (GTP) currently underway, can seem like small change [14].

The challenge, though, has been to orient the world of finance toward consistently adding value to the real economy in ways that are noticeable to average citizens. Put differently, where is the exit door from the casino, leading back to the real world? Innovation requires investment beyond seed capital. This begs the question how we can fund the upgrade needed to the market economy, and get financial markets to work for the real economy in ways that are sufficiently profitable. Over the past decade, a powerful idea has gained more and more traction: the simultaneous and intentional creation of economic and social or environmental value. The assumption is that this can be achieved in all sectors of the economy and society. Powered by innovative ways to access and allocate capital, incentives to achieve financial and extra financial performance simultaneously benefit shareholders and stakeholders alike. The quest for holistic value opens the path to use capital much more efficiently to fund solutions to social and environmental challenges.

Part IV asks how we can tap capital markets for new investment content that combines profits with impact, and assesses the potential contribution of a new show in town called "impact investing." Impact investing (also referred to as "social finance," "social impact investing," and "blended value investing" or "impact finance") has gained significant momentum in recent years, and there are numerous related ideas and terms, e.g., "sustainable finance." As we move through the first quarter of the twenty-first century, this kind of thinking could be one of the biggest ideas to renew the relevance of finance for the real economy and social progress. The market for impact investment holds great promise as a tool to grow the economy and fund the provision of public goods [15]. However, there is still a road to be travelled to grow to a critical mass in terms of market size. Given their current pace of building scale, delivering on this potential will require reaching outside the current networks of self-identified impact investors so that impact investments become a part of standard asset allocations in investor portfolios. Only with other supporting measures, creativity, and bold action can a functioning large-scale impact capital market emerge. The potential is huge. To date, only its surface has been scratched. We cannot afford an ivory tower or interest group led approach unlikely to bring it to fruition.

Part IV enquires into taking impact investing to a critical mass where it can make a noticeable contribution in the socioeconomic trajectories of countries in terms of capital formation, business opportunities, new jobs, and inclusive growth. It argues that this requires cleverly navigating new developments in finance resulting from the digital revolution, regulation, and the rise of Asia. The chapters discuss sources of fresh investment content in the areas of publicly listed companies, emerging markets, and the impact of technology on fresh opportunities resulting from the emerging Internet of Things (IoT). They build on insights stemming from work commissioned by the UK Cabinet Office (CO), and a baseline report I authored for the participants of the first-ever G8 conference on social investment held in 2013 under the UK presidency of the G8 in that year [16, 17]. Given the ambition to make a sizable contribution and the corresponding need to take the sector from niche to mega, we need to formulate a view regarding intelligent private sector and policy action to build up the impact investing market—or however else you want to call for-profit investments that consciously seek to achieve social and environmental performance as well.

Human Nature, a Constant in a Changing World

We live in times of dizzying change—in terms of technology, interconnectedness and mobility, and shifting social norms. If you see yourself as a member of Generation Impact, you are embracing the possibilities. But even if you are skeptical, it is worthwhile finding out why more and more people are building on some construction site of the Impact Economy.

To end up in the "yea" camp, a key challenge now is to find ways to leverage the new opportunities, inventions and innovations to drive the greater good and long-term viability of our species and the planet.

If you have made it through the building blocks discussed earlier—business innovation, sustainable supply chains, the future of energy, and inclusive capital markets—you might be interested in how all of this can fit together. How we can move onto the fast lane by understanding how human nature works? With the Factory 4.0 around the corner, about half of all "employees" in the European Union (EU) may be robots soon [18]. Notwithstanding, human nature is—for now—a constant in our changing world, as well as our great asset in driving progress and designing change, as well as the focus of the Epilogue.

References

1. Simmons M (2013) If you want to go fast, go alone. If you want to go far, go together. In: Forbes. http://www.forbes.com/sites/michaelsimmons/2013/07/22/power-of-relational-thinking. Accessed 2 Sept 2014
2. Meadows DH et al (1972) The limits to growth. Universe Books, New York
3. Credit Suisse (2014) Global wealth report. http://economics.uwo.ca/people/davies_docs/credit-suisse-global-wealth-report-2014.pdf. Accessed 1 May 2015
4. Global Impact Investing Network (2014) The GIIN register. http://www.thegiin.org/cgi-bin/iowa/resources/about/index.html. Accessed 11 July 2014
5. Cardone G (2011) The 10X rule: the only difference between success and failure, 6th edn. Wiley, New Jersey
6. Machiavelli N (1515) Il principe. http://www.constitution.org/mac/prince.pdf. Accessed 15 July 2015
7. LOHAS Groupsite (2014) The LOHAS Groupsite register. http://lohas.groupsite.com/main/summary. Accessed 6 Mar 2014
8. World Business Council for Sustainable Development (2010) Vision 2050: executive summary. http://www.wbcsd.org/pages/edocument/edocumentdetails.aspx?id=219. Accessed 13 Aug 2014
9. McKinsey & Company (2012) Winning the $30 trillion decathlon. http://www.mckinsey.com/features/30_trillion_decathlon. Accessed 16 Aug 2014
10. Brandes Investment Partners (2012) Five compelling reasons to allocate to emerging markets. http://www.brandes.com/Documents/Publications/Five%20Compelling%20Reasons%20to%20Allocate%20to%20EM.pdf. Accessed 16 Aug 2014
11. MarketLine (2012) Global textiles, apparel & luxury goods. http://store.marketline.com/Product/global_textiles_apparel_luxury_goods?productid=MLIP0468. Accessed 14 Nov 2013
12. Martin M (2013) Creating sustainable apparel value chains. A primer on industry transformation. Impact Econ. http://www.impacteconomy.com/papers/IE_PRIMER_DECEMBER2013_EN.pdf. Accessed 24 Aug 2014
13. Bastasch M (2014) Indian government promises enough solar energy to power a light bulb in every home. In: Daily Caller. http://dailycaller.com/2014/05/20/indian-government-promises-enough-solar-energy-to-power-a-light-bulb-in-every-home/. Accessed 20 May 2014
14. Ministry of Finance and Economic Development (2010) Growth and transformation plan 2010/11-2014/15, Volume I. http://www.mofed.gov.et/English/Resources/Documents/GTP%20English2.pdf. Accessed 28 Aug 2014

15. Global Impact Investing Network (2014) About impact investing. http://www.thegiin.org/cgi-bin/iowa/resources/about/index.html. Accessed 11 July 2014
16. Martin M (2013) Status of the social impact investing market: a primer. UK Cabinet Office. http://www.impacteconomy.com/de/primer1.php. Accessed 11 July 2014
17. Martin M (2013) Making impact investible. Impact Economy http://www.impacteconomy.com/en/wp4.php. Accessed 11 July 2014
18. Berger R (2014) Industry 4.0: the new industrial revolution: How Europe will succeed. https://www.rolandberger.com/media/pdf/Roland_Berger_TAB_Industry_4_0_20140403.pdf. Accessed 29 May 2015

Part I
Big Business: Climb to Sustainability?

Willingness to change is a strength, even if it means plunging part of the company into total confusion for a while.

Jack Welch, n.d.[1]

[1]Special Dictionary (n.d.) Famous quotes, Jack Welch. http://www.special-dictionary.com/quotes/authors/j/jack_welch/171889.htm#sthash.TjNOTosu.dpuf. Accessed 10 July 2015.

Brave New World—Sustainable Mega Value Trends

1

1.1 The Four Drivers of Sustainable Value Creation

Our world has changed dramatically since the advent of rapid industrialization in the West. Energy and growing consumption of the world's fossil fuels have powered global economic progress for the past three hundred years. Yet with the emergence of globalization, a resurgent Asia, and an unprecedented surge in wealth creation, humanity will soon test the limits of the planet as a whole for the first time in history. Many social problems remain unresolved, merely growing in scale and urgency. Innovation on a massive scale is now required to respond to the planet's growing population and shrinking resources—otherwise we might find ourselves in the "overshoot and collapse" scenario the Club of Rome prognosticated more than forty years ago sooner than we think.

The question is which role companies will ultimately end up playing in all of this. Modern capitalism has brought about unprecedented wealth and comfort. If the ascent to sustainability ends up being successful, corporations will be the good guys. If it fails, and the balance will be greenwashing, many of their products and services will have ultimately distracted us from the challenges we had better mastered. Then companies will be more like the Titanic's orchestra—keeping us entertained, moving around the deck chairs, while the ship goes under.

To succeed, sustainability thinking needs to move from the margins to the center, and fast. A transformative new path of sorts is needed, responding to structural changes in businesses' operating environment. This road has to be far more ambitious in terms of scale and impact, and endeavor to achieve actual financial returns, greater effectiveness, and more leverage potential for both business and society. Pioneers of this new approach are not just seeking to create a new corporate volunteering program or foundation. Nor are they simply building more public infrastructure and throwing proven models at new spatial frontiers to engineer long-term growth and prosperity. This was done in the post-World War II period, which led to the rapid industrialization of countries like Russia and China.

M. Martin, *Building the Impact Economy*,
DOI 10.1007/978-3-319-25604-7_1

Instead, by marrying impact, investment, and innovation, companies need to build what will be the next steam engine, railroad system, and Internet.

The current transformation of the corporate value chain is coinciding with broader demand and supply shifts around the world. This should come as no surprise. Growing demand at the Base of the Pyramid, consumers increasingly looking to gain meaning from their consumption, the reconfiguration of growth to be "green," and a welfare state increasingly at odds with fiscal and demographic realities have collectively forced a thoughtful reconsideration of the role of enterprise in achieving social and environmental impact.

These trends are also challenging traditional notions of corporate social responsibility. Instead, they emphasize opportunity. They are pushing companies to fully integrate sustainability into their business models. This drives a convergence of business and the social sector. Moreover, government is increasingly playing an important role in shaping the necessary conditions for developing new business frameworks (Fig. 1.1). CSR was once an externally driven activity. Now it could give way to value creation opportunities for all stakeholders, including shareholders. Companies need to find ways to derive economic value from the megatrends reshaping the economy and society. They need to do it in order to stay in the game long-term.

Let us first examine which megatrends are currently reshaping value creation. Next, we will come up with a framework for how companies can achieve significant financial gains by addressing social and environmental issues through core business practices. Sourcing the corresponding innovations and business models is a major challenge, especially for well-established companies. Rerouting a supertanker to new phenomena such as generation Y, disruptive business models for sustainable value creation, or creating a space for social intrapreneurs is often difficult. In many companies, the need to change breeds an atmosphere of fear, not creativity. But there is good news as well: business leaders do not have to have all the ideas themselves. Venture capital (VC) provides a helpful framework for how to invest in entrepreneurs with new ideas, models and technologies: startup entrepreneurs and scientists provide the ideas, and venture capitalists provide the funds. In the new supertanker setting discussed here, corporations can provide the supply chains and market access, as well as some of the capital. This enables future-oriented business ideas to take off and create economic, social and environmental value simultaneously.

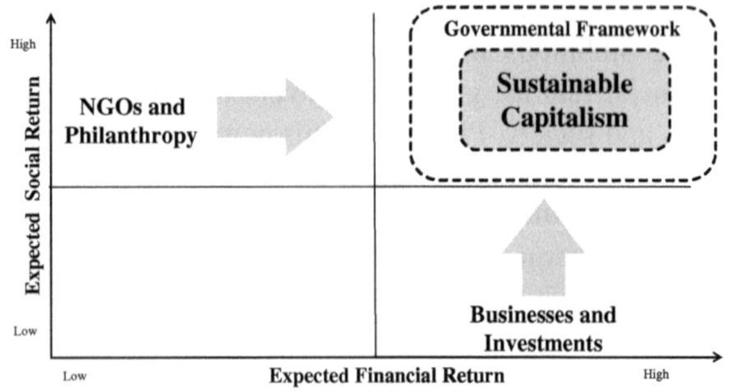

Fig. 1.1 The social-financial return matrix (*source* author) [1, p. 3]

Table 1.1 The four drivers of sustainable value creation

Theme	Rationale	Potential
Massive pent-up demand at the BoP	The BoP refers to the 4+ billion people with incomes below USD 1500 annually	The segment will take off as BoP consumers and producers join the market, already representing a USD 5 trillion economy
Emerging LOHAS segment at the ToP	LOHAS consumers prefer products designed to be environmentally conscious, sustainable, socially responsible, and better both for people and the planet	LOHAS consumption is already a USD 546 billion+ market worldwide and growing 10 % a year.
Driving green growth	The green economy results in improved human well-being and social equity, while significantly reducing environmental risks and ecological scarcities	Investments in clean energy assets including power, biofuels, and heat are forecast to increase from USD 195 billion in 2010 to USD 630 billion by 2030
Modernization toward a "triple E" welfare state	The welfare state needs to be modernized as some governments' expenditure is structurally higher than their revenues, with a large proportion of expenditure on healthcare, education and welfare	Fundamental shifts in the way we approach public good output lead to opportunities for the private sector, e.g. knowing that in 2020, 20 % of the worldwide population will be over 65

Source author [2, p. 4]

We are finally seeing the effects of social trends that have been bubbling just underneath the surface of our daily lives. Consider the increased demand of the poor; an aging population in most OECD countries, alongside a growing youth population in many emerging countries such as Iran; changes in green growth and total resource productivity; and a bifurcating consumer market driven only partially by price. These trends are pushing the world into uncharted territory. There are four "stories" that offer unprecedented possibilities to achieve large-scale value creation (Table 1.1). Each trend poses great adaptive challenges, but also distinct opportunities. Delivering on these opportunities can help us master the larger challenge of modernizing our businesses and economies so they can create value without destroying the environment.

1.2 Serve Massive Pent-up Demand at the Base of the Pyramid

Significant gains have been made in alleviating poverty, yet around 1.3 billion people live on less than USD 1.25 a day—the common cut-off for extreme poverty [3]. 805 million go hungry [4]. These populations are spread around the world, even in countries that have grown into "middle income" designations [5]. Much of this change is a result of countries like China and India moving into the global middle class [6]. Moreover, patterns of persistent exclusion from the economic mainstream in OECD countries mean that poverty is not the exclusive domain of the developing world.

The big market opportunity for companies is to serve the pent-up demand at the BoP—the roughly 4 billion people with incomes of less than USD 1500 annually—and help customers to graduate to the emerging middle class [7]. The BoP has been steadily gaining a heightened level of attention by business, investors, and development agencies.[1] If we do not find environmentally friendly ways to serve this demand, the result could be catastrophic, but let's map latent demand first.

Many emerging markets are already developing the fast-growing middle classes that have historically played a critical role in the development of major economies, whether in nineteenth-century Europe or twentieth-century America. The middle class is often thought of as the source of entrepreneurship and innovation—containing the small businesses that make a modern economy thrive. Middle class values also emphasize education, hard work and thrift. Thus, the middle class is an important source of the needed inputs for growth in an economy: new ideas, physical capital accrual, and human capital accumulation.

The focus on the BoP market will only grow because, according to Fortune researchers, the aggregate purchasing power of consumers at the BoP "is enormous, creating major opportunities for businesses that can deliver products they need at prices they can afford" [9]. Both for the current members of the Global 500 and those that aim to join it, many of their next customers are likely to come from the BoP, which is an estimated United States Dollar (USD) 5 trillion global market. Many companies that have historically served emerging markets are gradually joining the Global 500 or are at least growing very fast; Huawei from China or Mobile Telephone Networks (MTN), which is Africa's largest wireless carrier, are prime examples of this development.

In the area of telecom specifically, growth has exploded at the BoP. In 2014, there were about 7 billion mobile phone subscriptions, an average of almost one for every person on Earth [10]. This is equivalent to 95.5 % of the world population and that's a billion extra mobile subscriptions in the past three years. With 1.5 billion in the developed world, mobile phone subscriptions are rapidly reaching saturation point. Market growth is being driven by demand from the developing world, led by rapid mobile adoption in China and India, the world's most populous nations. There are 5.4 billion mobile subscriptions in the developing world—that's 78 % of global subscriptions (including prepaid contracts). Mobile penetration in the developing world now is 90 %, but there is still potential for growth, particularly in Africa, which has the lowest mobile penetration worldwide at 69 % [11]. The Asia-Pacific region today accounts for half the global smartphone connections total, despite the fact that penetration is still below 40 %. Smartphone penetration in sub-Saharan Africa is only around 15 % of mobile connections today [12]. There is tremendous opportunity to harness this development and serve both the consumer and businessperson through mobile innovations.

---

[1]Following the definition of the OECD, the global middle class can be defined as having a daily per capita income of between USD 10 and USD 100 in purchasing power parity terms. The middle class in the twelve leading emerging markets includes almost two billion people, spending a total of USD 6.9 trillion annually, roughly double the consumption of the United States [8].

While many in the developing world are using mobile phones, fewer are using the Internet. Most annual growth of Internet users still comes from the developed world. 30 % of the developing world is now online, compared to about 70 % of the developed world. 4.3 billion people are still not online, and 90 % of them live in the developing world [13]. However, the number of online users in the developing world is rapidly growing. Most of this growth comes from China, which makes up around 40 % of Internet users in the developing world [14]. Rather than connecting to the Internet from personal computers, emerging market Internet use is often mobile-enabled. This renders Internet usage growth patterns different from earlier adoption in the West.

There is still more demand to be met, which is a huge opportunity for businesses looking to serve the poor. The telecom example is but one of many. Investments in water access, sustainable forestry and fisheries, and civic infrastructure in developing countries will have significant returns both financially and socially [15, 16]. But to serve these markets, companies need to understand what matters to their consumers, the business models that work in practice, how to make distribution happen, and how to become a trusted product and service provider. Typically, this requires identifying pioneer business models and taking time to learn from implementation before attempting to scale. This approach has been the focus of several companies and foundations, such as the Shell Foundation, that are actively identifying and supporting emerging social businesses [17]. In some cases, collaboration between industry players makes it easier to "crack" BoP markets, and entails additional benefits for the core business, as in the case of Sanofi Group, described below (Spotlight 1).

Spotlight 1: Sanofi—"No Profit, No Loss" Drugs at the BoP

The creation of the Nifurtimox-Eflornithine Combination Therapy (NECT) drug, and the "no profit, no loss" approach taken by the Sanofi Group when it partnered for its creation illustrates the market opportunity at the BoP [18].

The development of the NECT was made possible by the cooperation of multiple actors from different sectors towards a social goal: combating "Human African trypanosomiasis" (also known as "sleeping sickness," a deadly parasitic disease that affects the nervous system of those that become infected) [19].

Developing drugs and accessing new markets is a costly endeavor for major pharmaceutical groups. To solve this problem, Sanofi defied conventional practices and in 2009 started a series of collaborations both with competitors and international organizations. Through a product development partnership (PDP) agreement, Sanofi was able to co-develop the drug and make it accessible to those that need it the most. Next to Sanofi, the PDP included Drugs for Neglected Disease initiative (DNDi), which developed and financed the clinical study, a business partnership with competitor Bayer Schering Pharma, which just like Sanofi, donated the nifurtimox and eflornithine drugs, and the World Health Organization (WHO), which supplied a free-of charge medical kit to patients [20, 21]. In 2011, 731 million people were treated for neglected tropical diseases (including Human African trypanosomiasis). This magnitude could never have been reached without the cooperation of the public, private and non-profit sectors under the "no profit, no loss" approach [22].

The return to Sanofi—and the underlying business rationale—was that the firm was able to gain "access to emerging markets where it otherwise had no presence (and where the company can eventually sell other drugs at higher margins), new knowledge of diseases that could make the leap to the developed world (through climate change or migration); and, of course, some brand polishing" [22].

1.3 Leverage LOHAS, a Pioneer Segment at the Top

Consumers are playing a major role in the emerging landscape. Once their basic needs are met, consumers increasingly look to gain meaning from their consumption. Conscious consumption, once on the fringes of society, is now rapidly moving into the mainstream. Consumers in this segment are sometimes called LOHAS for "Lifestyles of Health and Sustainability." They are realizing that they have the power to positively or negatively influence society with their purchases. More and more individuals want to create a lifestyle that fits with their values. Some industries have many actors while others have little LOHAS activity, but interest, new products and action are rising across the board. The push for eco-friendly and conflict-free smartphones and fashion products illustrates this interest. Since we will need massive amounts of less resource intensive and environmentally damaging products to serve demand at the Base of the Pyramid without destroying the environment in emerging markets, such innovations are of special interest. The rise of brands like (Product) RED, TOMS Shoes, fair trade labels such as Max Havelaar, and organic food retailers such as Whole Foods Market show how LOHAS businesses can serve consumers.

As the global middle class expands, from 1.7 billion people today to 3.6 billion in 2030, many more consumers will be reaching for sustainable products [23]. A survey of 8000 consumers in 16 markets found that interest in sustainability through consumerism continues to rise. Since 2008, 86 % of consumers report that they believe companies should focus as much on social impact as they do on their core business responsibilities. From 2008 to 2012, the number of consumers stating that it is acceptable for companies to support good causes and make money at the same time increased by a third. Nearly three-quarters of consumers will recommend a brand that supports a good cause over one that does not, up almost 40 % since 2008. About three-quarters also said they would switch products if another with similar quality supported a good cause [24]. Many consumers are interested in businesses engaging in "products with impact" strategies: instead of simply donating to a charity to create impact, almost 50 % of consumers want businesses to create new products that solve societal problems [25].

While many people express interest in buying sustainably, consulting firm Deloitte found that only about 20 % actually end up purchasing a sustainable product. The Deloitte study concluded that the disconnect between what people say they want in a product, as opposed to what they actually purchase, pointed to significant market growth potential in the LOHAS segment [26]. However, pricing also plays a major influencing factor here. The growth potential is tremendous but conditional to large extent on equal price. Many consumers are willing to consume more sustainably but not at a higher price. This condition is unlikely to be relaxed in the low growth context of aging advanced economies.

These trends are not limited to the developed world. While reliable data is scarce, some studies indicate just the opposite. Middle-class consumers in middle-income countries such as China, Brazil, India, Indonesia, Malaysia as well as the United Arab Emirates are more likely to be inclined towards companies and products with a purpose than their advanced economy counterparts [26]. The proliferation of the sustainable consumer in these emerging markets is supposedly so strong that they have been labeled "Purpose Bull Markets." According to the study, the consumers in these "bull markets" were much more likely to act on their interest in sustainable products; 62 % of "bull" consumers made a sustainable purchase monthly, compared to 37 % elsewhere [26]. This astounding finding is so counterintuitive that it raises eyebrows in strategy sessions.

In Eastern Europe and Central Asia, sustainable consumption takes a different form. The issues do not arise from patterns of overconsumption. Instead, sustainable consumption "may need to be more weighted towards improving efficiencies of production, infrastructures and municipal services" in these regions [27]. While the data points will multiply in the years to come, there is already enough evidence to suggest that companies should consider the LOHAS trend and think hard about business innovations in their value chains. To illustrate the opportunity, consider the story of retailer Whole Foods Market, as described below (Spotlight 2).

Spotlight 2: Whole Foods—Shopping Food to Make a Statement

The LOHAS segment is growing. A market scan by the Natural Marketing Institute (NMI) from 2008, estimated that US Consumers spent roughly USD 300 billion on LOHAS-related products. This figure represented a growth in spending of 16 % from 2005, the prior year that similar data was surveyed. Six market segments generated most of the growth, including green building, personal health, eco-tourism, alternative transportation, natural lifestyles, and alternative energy [28]. More recent figures suggest a global market penetration of USD 546 billion worldwide [29].

A helpful illustration of the magnitude of opportunity for business transformation in the LOHAS segment is Whole Foods Market, the leading chain of natural foods supermarkets in the US. Whole Foods has become the poster child, perhaps more so than any other brand. Starting in 1978 with one store (then known as "SaferWay", not to be confounded with "Safeway"), the company has greatly expanded over the years, with sales of USD 12.9 billion and operating income of USD 883 million in 2013 [30].

Whole Foods approaches its work with a combination of activities: it rewards its employees, emphasizes autonomy and creativity, pays executives reasonably, sets animal welfare standards, gives preference to artisan producers, makes strong environmental commitments, and uses fully renewable power.

The result is that people see Whole Foods more as a lifestyle store, as opposed to just another grocery store. "Some customers say they are making a statement by shopping there. Their motto 'Whole Foods, Whole People, Whole Planet' emphasizes the company's vision as more than just a food retailer" [31]. Other leading brands—from Google, to Amazon, Starbucks, Panera Bread, and UPS—have taken note [32].

The mainstreaming of LOHAS becomes clear when looking at the top five organic food retailers in the US: Walmart, Costco, Kroger, SuperTarget and Safeway [33]. All are incumbent food retail businesses that have discovered the opportunity for sustainable value creation in their core businesses, also offering organic food.

1.4 Boost Total Resource Productivity

As humanity is about to test the ecological limits of the planet as a whole for the first time in history, we have entered uncharted territory as to where the consequences of the resource hungry Industrial Revolution will take human civilization. This has important implications for value creation in business far beyond themes such as green energy. On our current carbon emissions trajectory, the world is set to a temperature rise exceeding four degrees Celsius by the end of the century. Temperature rises are forecast to be even more pronounced along the Polar Regions and new shipping routes through the Arctic are already being planned. For many, a sustainable future is not yet in sight. A report published by the World Business Council for Sustainable Development (WBCSD) estimated that if we continue on our current consumption path, we would require the equivalent of 2.3 Earths to survive in 2050 [23].

The *how-to* question is: what can we do about it? This presents a major challenge and a need to shift our fundamental economic infrastructure. Yet again, this challenge is also an opportunity. Access to energy is and will remain a fundamental human and economic development need. We know the transition to a low-carbon economy will impact all industries either directly or indirectly [34]. The business opportunities of the future will not necessarily resemble or follow the same patterns of the past. Regardless of whether humanity achieves reduction goals in greenhouse gas (GHG) emissions, the scarcity of natural resources will require investments in new infrastructure in order to meet the needs of a growing population.

The WBCSD report estimated that the sustainability-related industries of natural resources, health, and education (the latter two comprising of our planet's "social sustainability") could reach USD 3-10 trillion annually by 2015, in 2008 dollars [23]. Areas such as water and mining waste are forecast to become the new industries of the future. Water could require annual investments of up to USD 200 billion until the year 2030. Systematic waste management, or the transport, treatment and disposal of waste, will become the new normal [35]. WBCSD estimates that the opportunity to re-capture aluminum foil alone is USD 5.6 billion [23]. Next to investments in green growth, human capital will become important as well: health and education, in both developed and developing countries will provide some of the most attractive financial and social returns. As detailed below (Spotlight 3), significant transformation across the board can be expected by the gradual transition to a circular economy, where resources are not discarded after use.

There is also a major opportunity in renewables, discussed more fully in Part III on energy. Due to shifting regulation, technology risk, and volatile commodity prices, many investors have lost money in green investments in the past. But to truly mitigate the effects of global climate change, significantly higher investments in renewable energies must be made [36]. Investments in renewable energies are forecast to grow significantly, rising threefold from USD 195 billion in 2010 to USD 630 billion 2030 in nominal terms [37]. Growth in renewable energies is occurring in developing countries, such as Brazil, India, and China. These growing economies need affordable long-term reliable energy supplies to thrive [38]. A key question is how governments can unlock much needed private sector investment to upgrade the physical infrastructure to facilitate green growth. This brings us quickly to the role capital markets can play to finance the Impact Economy, the subject of Part IV. Accelerating capital deployment will require incentives such as public sector entities absorbing some green investment risk via first loss provisions.

Conducive frameworks are equally needed, and this means intelligent planning to envision and design paths for how to make the transformation to green infrastructure happen concretely at the local, regional and national levels. For example, service and residential buildings are currently responsible for 39.6 % of energy consumption in Europe, ahead of transportation and the industrial sector [39]. Ultimately, renewables can only succeed if they can offer full substitution of fossil fuels in terms of price and performance. In Part III, we will look at the technical hurdles still standing in the way, and how to deal with them.

Spotlight 3: Circular Economy—Resuscitation versus Cradle to Grave

Economic progress over the past three hundred years has been driven by a linear logic: economic development came largely as a function of the one-time use of a seemingly never-ending supply of additional factors of production. This model served humanity well until the mid-twentieth century, when the finite nature of natural resources and the impact of human activity on the planet came to the forefront. Looking ahead, estimates indicate that from 2010 to 2025, food caloric consumption could increase by 24 %, food spending by 57 %, packaging by 47 %, and end-of-life materials by 41 % [40]. With a current population growth scenario of 9 billion by 2050, the outlook is not promising if we continue business as usual.

In response to growing resource constraints, production in a variety of industries is now gradually shifting to a circular model—from cradle to cradle, instead of cradle to grave. Attention is increasingly focusing on the total resource productivity of the factors of production through materials, product and process innovation. Other areas are avoiding unnecessary waste in packaging, reusing waste and extending the product lifespan. Recycling rates are already fairly high for some common materials in the major economies, whereas they are low for others. In the European Union (EU) and China, 90–95 % of aluminum used in the building and transport sectors is already being recycled, but only 24 % of plastic wastes in the EU (compared to 22 % of plastics in China). By 2015, 15 % of total manufactured fibers are expected to be recycled in China, and one quarter in the EU [41–46]. These rates are rising and new materials, such as polyester made from recycled materials rather than corn, will take a larger market share in due course.

The US has traveled partially down this road. In 2012, 55 % of aluminum beer and soft drink containers were recycled, 9 % of total plastic waste and 15.7 % of total textiles [47]. The Ellen MacArthur Foundation argues that fast moving consumer goods (FMCG) industries can in the "near term" move from a current model, where 80 % of consumer goods are not recovered to a model where non-recovery drops to 50 % [48]. In this scenario, transitioning to a circular setup in relevant fast-moving consumer goods sectors could yield net material cost savings of USD 595–706 billion per year at a global level [48, p. 56]. More recent figures from follow-up research, which the Foundation released in 2014, indicate an even greater savings potential of USD 1 trillion a year once non-fast moving consumer goods are added to the mix [49].

The speed of the mainstreaming of the transition to more circular business models remains unclear. But players in a variety of industries are seriously examining how to integrate circularity into their business models. A number of well-known companies are members of the "Circular Economy 100," which is a group convened by the Ellen MacArthur Foundation bringing together companies, innovators and regions to accelerate their transition to a circular economy [50]. Moving toward circularity will alter the cost structures of firms. This has the potential to create a comparative advantage if resource prices rise substantially, or if access to resources becomes limited. Once successful, such a dynamic will force other players to adapt as well. In practice, however, the effective transitioning of a firm to a circular economic model is difficult to envision without the onboarding of significant outside expertise and possibly making acquisitions that bring the circular business focus and technologies needed.

Another opportunity to reduce the CO_2 footprint comes from more efficient supply chains. More transparent supply chains are an important first step towards achieving this goal. Consider the case of Puma, the German sportswear manufacturer. Together with Kering (formerly known as Pinault-Printemps-La Redoute or PPR), which is Puma's French parent company, Puma published an environmental profit and loss account for the first time in 2011 [51]. Puma was the first clothing and footwear group worldwide to publish such figures. Raw material extraction and production represented the largest share of greenhouse gas emissions and water consumption within Puma's core business and its supply chain. The direct costs of the environmental impact of greenhouse gas emissions and water consumption by the company was approximately EUR 7.2 million. A further EUR 87.2 million of environmental impact fell along Puma's four supply chain levels, adding up to EUR 94.4 million in total [51].

Some countries with historical growth models that were reliant on economies of scale and fossil energy-led industrialization are still in relatively early stages of seizing the opportunities of green growth. Russia achieved amazing economic growth from 1900 to 2000, with its economy growing 36.7-fold [52]. However, Russia has so far seen little movement in renewable energy technologies. Developments are limited to small-scale projects. With Russia's vast capacity for hydro, biofuel, and wind energy in particular, there is large potential to tap into the renewable market for increased financial and environmental returns [53]. Across Eastern Europe and Central Asia, energy production still suffers from an overall efficiency gap compared to EU member states, with significant technological advances needed to improve the process [53, p. 19]. While reducing greenhouse gas emissions has thus far not been achieved, underestimating the consequences of the many efforts focused on this issue would be shortsighted. Before we look at breakthrough innovations in clean energy in Part III, consider the case of GE Ecomagination below (Spotlight 4).

Moreover, there are many ramifications of climate change and population growth, including the huge need for better agricultural productivity. To satisfy the needs of future population, the world agricultural production would need to increase by about 60 % (77 % for developing countries alone) between 2005/06 and 2050 [54]. This requires investments of approximately USD 83 billion each year, to a large extent in developing countries [23]. In the prognosticated worst-case scenario, the world would have 39 % less harvested areas in 2080. Developing countries would have 29 % less, which might result in an additional 130 million undernourished people in sub-Saharan Africa [23].

Spotlight 4: GE—Does Greener Technology Do the Trick?
Launched in 2005, GE's Ecomagination program seeks to achieve advancements in five major areas of activity, including: doubling investment in cleantech research and development (R&D), increasing revenues from Ecomagination products, reducing greenhouse gas (GHG) emissions and improving the energy efficiency of operations, reducing water use and improving water reuse, and keeping the public informed [55].

According to a 2011 Ecomagination progress report, the overall portfolio had generated more than USD 105 billion in sales and revenues from a series of new products and services like the FlexEfficiency*50 Combined Cycle Power Plant. The FlexEfficiency*50 delivers "cleaner, more efficient energy onto the power grid and into […] homes" [56]—by enabling the "integration of more renewable resources onto the power grid by combining efficiency and flexibility to rapidly ramp up when the wind is not blowing or the sun is not shining, and to efficiently ramp down when they are available" [57]. Sounds great, does it not?

Operationally, the FlexEfficiency*50 plant creates power far more efficiently than current power plants (61 % more efficiently) and has a ramp-rate of "more than 50 megawatts (MW) per minute, twice the ramp-rate of today's industry benchmarks" [57, p. 25]. The plant can be "integrated with a concentrated solar power field to achieve more than 70 % efficiency and it offers a 10 % smaller footprint than existing combined cycle power plants with equivalent output" [57, p. 25].

The environmental and operating benefits of the FlexEfficiency*50 are significant. The average power plant could save USD 2.6 million, annual fuel savings of 6.4 million cubic meters of natural gas (the equivalent of 4000 EU households), and would reduce CO_2 emissions yearly by more than 12,700 metric tons (the equivalent of removing more than 6000 cars) [57, p. 26].

However, energy is a highly regulated and complicated industry. Whether additional efficiency and flexibility is ultimately enough to compete as the cards between renewables and conventional power plants are being redistributed remains to be seen. To get a flavor, consider the saga of the Irsching power station in Bavaria [58]. As Germany increases its reliance on volatile generation sources such as solar and wind, the government is paying some plants to guarantee stable power supply under a so-called redispatch agreement. Irsching is a top-notch high-efficiency billion Euro+ combined-cycle power plant with an efficiency of 60 %. Yet, Irsching is one of many power stations in Germany that are running at a fraction of the time needed to be profitable, replaced by solar and wind energy, which has priority access to the grid. Overall, gas-fed plants are struggling to compete with renewables and coal-fired facilities in Germany. In May 2015, E.ON was losing EUR 16.03 per Megawatt-hour (the "spark spread"). However, between April 2013 and March 2016, the grid operator pays a share of the fixed costs of Irsching Units 4 and 5 (an annual double-digit million-Euro payment for each block). In turn, E.ON agreed to make the facilities available for generation for three years. That agreement has been called a one-off. Unsurprisingly, on March 30, 2015 E.ON declared they wanted to close down all of its operations effective April 1, 2016. This decision put pressure on Bavaria's state government, which has called for Irsching to continue operating. As Bavaria blocks links to transport renewable power from its northern states, Germany is now considering splitting its electricity market into two zones. This would probably increase prices in the southern state. We will look into renewable energies in detail in Part III.

1.5 Unleash a Triple-E Revolution on the Welfare State

The fourth key driver stems from the current modernization of the welfare state. The shifting demographics of the world and severe fiscal realities have combined to form our final trend. Major new demand for products and services is around the corner in areas that were once the near-exclusive purview of OECD governments, such as healthcare, education, physical infrastructure, unemployment, and security. What's needed is a "triple E" perspective: greater efficiency, greater effectiveness, and greater engagement with citizens to allow for intelligent subsidiarity. A mere focus on slashing costs will not provide the room needed to innovate to actually address the deeper issues, such as the lack of financing for prevention. But we also do need to raise the efficiency of the overall system, because entitlements cannot simply keep growing to a point where they bankrupt nations. Leaving space for citizens to engage provides meaning and saves taxpayer money. Those not paying attention as the opportunities form are at risk to miss out. Let's unpack this.

Many factors at play will leave the public sector unable to provide for today's social welfare state. At the top of the list are fiscal concerns: the welfare state is relying on uncertain economics, obsolete demographic models, and is no longer affordable. The US Medicare system, which provides health insurance to retired workers, is estimated to be bankrupt as soon as 2021, absent a major overhaul. To make up the gap in payments to revenues, the system would need to collect more than USD 30 trillion [59]. A fundamental debate is arising around the entitlements versus (vs.) the costs of aging. Medicines and medical technologies help us to grow older, but at significant and increasingly unaffordable costs, coinciding with a discussion on quality of life. A rapidly aging global population will require fundamental shifts in the way we approach our economic output, as well as our social safety nets. In 2020, people over the age of 60 will make up more than 21 % of the population worldwide [60]. Just consider the impact of the recent low or negative interest rates on pensioners: without appropriate funding solution redesign, most people will ultimately have to work until they die.

In 2015, China decided to relax its one-child policy to counter the effects of an aging population on economic development [61]. Notwithstanding, in the short term, the modernization of the welfare state is primarily an advanced economy phenomenon. For all advanced economies in North America, Europe, East Asia and Oceania, the peak ratio of workers contributing to the welfare system to dependents —those benefiting from the welfare system—happened between 2000 and 2010 [62]. These countries must update their welfare states to support the increasing number of dependents. In developing countries, that peak ratio is further away. In South East Asia and Latin America, the number of dependents will start increasing around 2025; in Northern Africa, Western Africa, and South-Central Asia, it will not happen until around 2040. Sub-Saharan Africa has the longest time-horizon to build a future proof welfare state: its demographic shift will not occur until well after 2050 [62].

The welfare state as we know it may be coming to an end, or at least be drastically overhauled. Yet the issues that the welfare state was created to address are not. They are more likely set to get worse. The issues of poverty and inequality at all age levels will not disappear as governments around the world debate how to handle the needs of an aging population. More efficient and self-service prevention are a fast-growing market.

Here too, though, change is afoot as innovative developments by public and private actors have emerged in response, using capital markets to fund the adjustment, as discussed in Part IV. For example, Social Impact Bonds/pay for success schemes (SIBs) are increasingly gaining traction. A contingent return model that links a financial return to the achievement of a social outcome, SIBs "raise private investment capital to fund prevention and early intervention programs that reduce the need for expensive crisis responses and safety-net services" [63]. With this approach, "[if] improved outcomes are not achieved, the government is not required to repay the investors, thereby transferring the risk of funding prevention services to the private sector and ensuring accountability for taxpayer money" [63, pp. 4–5].

Models like SIBs are demonstrating new approaches to the provision of public goods that harness both private and public resources and focus on effectiveness. In due course, these approaches may reach a scale where they become an interesting product innovation opportunity for the financial services industry. A few citizens who chose to engage kicked off this development; philanthropic funding was instrumental to getting SIBs off the ground. One example of the power of engagement in shaping the triple E welfare state.

Adjusting from an exclusive focus on corporate responsibility to corporate opportunity in the social and environmental domains will be key to creating value from new demand patterns. Focusing on pathways that can turn up the fresh business ideas needed to turn demand into successful products and services will be critical. For corporations to contribute to solving the problems in a major way, we need to be entrepreneurial. To answer the *how-to* questions, we need to practice venturing. This is the focus of the chapters that follow.

References

1. Martin M (2014) Driving innovation through corporate impact venturing: a primer on business transformation, p 3. Impact Economy. http://www.impacteconomy.com/en/primer3.php. Accessed 15 July 2015
2. Martin M (2014) Driving innovation through corporate impact venturing: a primer on business transformation, p 4. Impact Economy. http://www.impacteconomy.com/en/primer3.php. Accessed 15 July 2015
3. World Bank (2015) Poverty overview. http://www.worldbank.org/en/topic/poverty/overview. Accessed 22 May 2015
4. FAO, IFAD and WFP (2014) The state of food insecurity in the world 2014. http://www.worldbank.org/en/topic/poverty/overview. Accessed 22 May 2015
5. Poverty Environment Partnership Joint Agency Paper (2012) Building the inclusive green economy for all. http://www.wri.org/publication/building-inclusive-green-economy-all. Accessed 25 Jan 2014
6. World Business Council for Sustainable Development (2010) Changing pace. http://www.wbcsd.org/Pages/EDocument/EDocumentDetails.aspx?ID=14622. Accessed 25 Jan 2014
7. Simanis E, Duke D (2014) Profits at the bottom of the pyramid. Harvard Bus Rev. https://hbr.org/2014/10/profits-at-the-bottom-of-the-pyramid. Accessed 22 May 2015
8. IFC (2007) The next 4 billion: market size and business strategy at the base of the pyramid, p 3. http://pdf.wri.org/n4b_full_text_lowrez.pdf. Accessed 6 Mar 2014
9. CNNMoney.com (2012) What will the global 500 look like in 2021? http://money.cnn.com/galleries/2012/news/companies/1207/gallery.global-500-future.fortune/index.html. Accessed 5 Mar 2014
10. Gillet J (2014) Measuring mobile penetration: Untangling subscribers, mobile phone users and users. In: GSMA intelligence. https://gsmaintelligence.com/research/2014/05/measuring-mobile-penetration/430. Accessed 22 May 2015
11. Mobithinking (2014) Global mobile statistics: 2014 part A: mobile subscribers; handset market share; mobile operators. http://mobiforge.com/research-analysis/global-mobile-statistics-2014-part-a-mobile-subscribers-handset-market-share-mobile-operators. Accessed 22 May 2015
12. Meyer D (2014) Emerging markets will shape the smartphone world, new industry figures suggest. In: GIGAOM mobile news. https://gigaom.com/2014/09/11/emerging-markets-will-shape-the-smartphone-world-new-industry-figures-suggest. Accessed 22 May 2015

13. ITU (2014) The World in 2014: ICT facts and figures. http://www.itu.int/en/ITU-D/Statistics/Documents/facts/ICTFactsFigures2014-e.pdf. Accessed 22 May 2015
14. ITU (2012) Measuring the information society. http://www.itu.int/ITU-D/ict/publications/idi/material/2012/MIS2012_without_Annex_4.pdf. Accessed 17 Feb 2014
15. McKinsey/Water Resources Group (2009) Charting our water future. http://mckinseyonsociety.com/charting-our-water-future. Accessed 6 Mar 2014
16. Poverty Environment Partnership Joint Agency Paper (2012) Building the inclusive green economy for all. http://www.wri.org/sites/default/files/pdf/Building_Inclusive_Green_Economy_for_All.pdf. Accessed 6 Mar 2014
17. Hystra (2012) Marketing innovative devices for the base of the pyramid. https://static.squarespace.com/static/51bef39fe4b010d205f84a92/t/51f237c4e4b07e4e5ac4e0f6/1374828484103/Full_report_Maketing_for_the_BOP.pdf. Accessed 17 Jan 2014
18. Warner B (2014) No profit, no loss, but a win against tropical disease. Bloomberg Businessweek. http://www.businessweek.com/articles/2013-01-23/no-profit-no-loss-but-a-win-against-tropical-disease. Accessed 14 Mar 2014
19. World Health Organization (2014) Human African trypanosomiasis. http://www.who.int/trypanosomiasis_african/en/. Accessed 14 Mar 2014
20. Drugs for Neglected Diseases initiative-DNDi (2014) ECT (HAT) = Nifurtimox-eflornithine combination therapy. http://www.dndi.org/treatments/nect-c-treatment. Accessed 14 Mar 2014
21. Franco JR et al (2012) Monitoring the use of nifurtimox-eflornithine combination therapy (NECT) in the treatment of second stage gambiense human African trypanosomiasis. In: WHO research and reports in tropical medicine. Dove Press, London. http://www.who.int/selection_medicines/committees/expert/19/applications/NE_6551_pub1.pdf. Accessed 14 Mar 2014
22. Warner B (2014) No profit, no loss, but a win against tropical disease. Bloomberg Businessweek. http://mobile.businessweek.com/articles/2013-01-23/no-profit-no-loss-but-a-win-against-tropical-disease. Accessed 13 Mar 2014
23. WBCSD (2010) Vision 2050: executive summary. http://www.wbcsd.org/pages/edocument/edocumentdetails.aspx?id=219. Accessed 25 Jan 2014
24. Cone C (2012) Introducing: good purpose 2012. http://purpose.edelman.com/slides/introducing-goodpurpose-2012. Accessed 25 Jan 2014
25. Edelman Insights (2012) Good purpose 2012: global findings. http://www.slideshare.net/secret/4FBiBICpOiDK3r. Accessed 21 Feb 2014
26. Deloitte (2009) Finding the green in today's shoppers. http://www.deloitte.com/assets/Dcom-Lebanon/Local%20Assets/Documents/Consumer%20Business/DeloitteGreenShopperStudy_2009.pdf. Accessed 2 Mar 2014
27. European Environment Agency-EEA (2007) Sustainable consumption and production in Southeast Europe and Eastern Europe, Caucasus and Central Asia. http://www.eea.europa.eu/publications/eea_report_2007_3. Accessed 3 Mar 2014
28. Dossey L (2010) LOHAS market size: U.S. consumer sales soar to close to $300 billion. Natural Marketing Institute, Spring, 2010. http://www.lohas.com/sites/default/files/lohasmarketsize.pdf. Accessed 3 Mar 2014
29. LOHAS Groupsite (2014) Summary. The LOHAS Groupsite Register. http://lohas.groupsite.com/main/summary. Accessed 6 Mar 2014
30. Whole Foods Market (2014) 2013 Annual report: financial highlights. http://www.wholefoodsmarket.com/sites/default/files/media/Global/Company%20Info/PDFs/WFM-2013-Annual-Stakeholders-Report.pdf. Accessed 5 Mar 2014
31. Harbin J, Humphrey P (2008) Whole Foods Market, Inc. J Case Res Econ. http://www.aabri.com/manuscripts/09288.pdf. Accessed 5 Mar 2014
32. Schiller B (2013) Is Whole Foods a viable model for the future of capitalism? FastCo. Exist. http://www.fastcoexist.com/1681352/is-whole-foods-a-viable-model-for-the-future-of-capitalism?partner=newsletter&goback=%2Egmp_4847080%2Egde_4847080_member_218268615. Accessed 3 Mar 2014

33. Chait J (2011) 6 largest organic retailers in North America 2011. http://organic.about.com/od/marketingpromotion/tp/6-Largest-Organic-Retailers-In-North-America-2011. Accessed 10 Jan 2014
34. Iltis C (1971) Leibniz and the vis viva controversy. Chicago J 62(1):21–35
35. OECD (2003) Glossary of statistical terms. https://stats.oecd.org/glossary/detail.asp?ID=2900. Accessed 6 Mar 2014
36. UNEP (2011) Green economy report: renewable energy, investing in energy and resource efficiency. http://www.unep.org/greeneconomy/Portals/88/documents/ger/6.0_RenewableEnergy.pdf. Accessed 6 Mar 2014
37. Turner G (2013) Global renewable energy market outlook 2013. Bloomberg New Energy Finance. http://about.bnef.com/fact-packs/global-renewable-energy-market-outlook-2013-fact-pack. Accessed 6 Mar 2014
38. UNEP (2011) Green economy report: renewable energy, investing in energy and resource efficiency. http://www.unep.org/greeneconomy/Portals/88/documents/ger/6.0_RenewableEnergy.pdf. Accessed 21 Feb 2014
39. EEA (2014) Final energy consumption by sector. http://www.eea.europa.eu/data-and-maps/daviz/total-final-energy-consumption-by-sector-1. Accessed 22 May 2015
40. Ellen MacArthur Foundation (2013) Towards the circular economy, 2: opportunities for the consumers goods sector. http://www.ellenmacarthurfoundation.org/business/reports. Accessed 9 Jan 2014
41. Organization of European Aluminum Association & Organization of European Aluminum Refiners and Remelters (2007) The Alu Europe register. http://www.alueurope.eu/wp-content/uploads/2011/08/Aluminium-recycling-in-Europe-2007.pdf. Accessed 6 Mar 2014
42. Plastics Recyclers Europe (2012) The plastics recyclers register. http://www.plasticsrecyclers.eu/sites/default/files/EuPR%20Strategy%20Paper%202012_0.pdf. Accessed 6 Mar 2014
43. European Association of Plastics Recycling & Recovery Organizations (2013) The Epro plastics recycling register. http://www.epro-plasticsrecycling.org/pages/75/epro_statistics. Accessed 6 Mar 2014
44. European Parliament (2014) Plastic waste: parliament sounds the alarm. http://www.europarl.europa.eu/news/en/news-room/content/20140110IPR32385/html/Plastic-waste-Parliament-sounds-the-alarm. Accessed 3 Mar 2014
45. Global Times (2011) Textile recycling. http://www.globaltimes.cn/NEWS/tabid/99/ID/662706/Textile-Recycling.asp. Accessed 6 Mar 2014
46. REdUSE (2014) Less is more. Resource efficiency through waste collection, recycling and reuse of aluminum, cotton and lithium in Europe, p 16. http://www.foeeurope.org/sites/default/files/publications/foee_report_-_less_is_more_0.pdf. Accessed 3 Mar 2014
47. United States Environmental Protection Agency (2012) Wastes—resource conservation—common wastes & materials: aluminum/plastics/textiles. http://www.epa.gov/epawaste/conserve/materials/alum.htm. Accessed 6 Mar 2014
48. Ellen MacArthur Foundation (2013) Towards the circular economy, 2: opportunities for the consumers goods sector, p 86. http://www.ellenmacarthurfoundation.org/business/reports. Accessed 9 Jan 2014
49. Ellen MacArthur Foundation (2014). Towards the circular economy, 3: accelerating the scale-up across global supply chains. http://www.ellenmacarthurfoundation.org/business/reports. Accessed 1 Mar 2014
50. Ellen MacArthur Foundation (2014) What is the circular economy 100. http://www.ellenmacarthurfoundation.org/business/ce100. Accessed 1 Mar 2014
51. Puma (2011) Puma and PPR announce first results of unprecedented environmental profit & loss account. http://about.puma.com/puma-and-ppr-home-announce-first-results-of-unprecedented-environmental-profit-loss-account/. Accessed 6 Mar 2014
52. Fedorenko NP, Simtchera VM (2002) Gross industrial production of Russia during the last 100 years. http://www.rus-stat.ru/eng/index.php?vid=1&year=2002&id=9. Accessed 15 Feb 2014

53. EEA (2007) Sustainable consumption and production in Southeast Europe and Eastern Europe Caucasus and Central Asia, p 58. http://www.eea.europa.eu/publications/eea_report_2007_3. Accessed 3 Mar 2014
54. Konuma H (2015) FAO regional office for Asia and the Pacific: policy on food crop and energy crop in the future. Paper presented at 16th TSAE national conference and 8th TSAE international conference Bangkok, 17 Mar 2015
55. Ecomagination GE (2014) Learn about ecomagination. http://www.ecomagination.com/portfolio. Accessed 11 Mar 2014
56. GE (2011) Progress. Ecomagination report 2011. http://files.gecompany.com/ecomagination/progress/GE_ecomagination_2011AnnualReport.pdf. Accessed 6 Mar 2014
57. GE (2011) FlexEfficiency*50, integrated renewable combined cycle power plant, eSolar announcement press kit. http://www.ge-flexibility.com/static/global-multimedia/flexibility/documents/FlexEfficiency_50_ePress_eSOLAR_kit.pdf. Accessed 6 Mar 2014
58. Frankfurter Allgemeine Zeitung (2015) Seehofer glaubt nicht ans aus für Irsching. http://www.faz.net/aktuell/wirtschaft/energiepolitik/modernes-gaskraftwerk-seehofer-glaubt-nicht-ans-aus-fuer-irsching-13523636.html. Accessed 21 July 2015
59. Roy A (2011) Saving medicare from itself. National Aff (Summer). http://www.nationalaffairs.com/publications/detail/saving-medicare-from-itself. Accessed 25 Jan 2014
60. United Nations Department of Economic and Social Affairs Population Division (2013) World population ageing 2013. http://www.un.org/en/development/desa/population/publications/pdf/ageing/WorldPopulationAgeing2013.pdf. Accessed 22 May 2015
61. GE (2011) FlexEfficiency*50, integrated renewable combined cycle power plant, eSolar announcement press kit. http://www.ge-flexibility.com/static/global. Accessed 6 Mar 2014
62. Bloom DE, Canning D (2011) Demographics and development policy. Harvard Program on the Global Demography of Aging. https://www.hsph.harvard.edu/pgda/WorkingPapers/2011/PGDA_WP_66.pdf. Accessed 3 Mar 2014
63. Social Finance (2012) A new tool for scaling impact: how social impact bonds can mobilize private capital to advance social good, pp 4–5. http://www.socialfinanceus.org/sites/socialfinanceus.org/files/small.SocialFinanceWPSingleFINAL.pdf. Accessed 16 Jan 2014

To Solve Problems, Take a Fresh View at Opportunity

2

2.1 Bye-Bye, CSR

"Sustainability" is a powerful idea that is leading to new opportunities, challenges and innovations across all sectors to develop a more prosperous future. In 2010, despite the then-recent economic downturn, an overwhelming majority of corporate CEOs in Accenture's research study on corporate sustainability—93 %—responded that sustainability will be critical to the future success of their companies. What's more, they believed that a tipping point could be reached that fully meshes sustainability with core business within a decade—fundamentally transforming core business capabilities, processes, and systems throughout global supply chains and subsidiaries [1].[1]

However, the actual response has been far more timid. The problem is that when companies are asked to care, their responses are usually formulated as part of the paradigm of a firm's corporate social responsibility (CSR). There are inherent limitations to current CSR practices that prevent businesses to respond in scale and fully capture the opportunities of the trends outlined above. Seen that an immediate complete overhaul of the CSR paradigm in large-scale and often politicized organizations would be unrealistic, this begs the question for the way forward. We need a clearer picture and a deeper understanding of current and future innovations, as well as a much more efficient way to source and rollout the business innovations needed. As we are entering a new phase of the market economy, the upside is to reconcile opportunity with responsibility. This creates both value and reduces risk. If more efficient and enlightened business practices can provide products and services to run with the trends discussed earlier, we can create both social and financial value on a scale not yet seen.

[1] Not all businesses operate globally or respond to such surveys. Many business entities serve local markets. However, global sustainability drivers such as resource scarcity or energy costs will similarly affect their ability to create value. In advanced economies and developing countries alike, the transformation of business currently underway will not be limited to just the cohort of large-scale global companies.

M. Martin, *Building the Impact Economy*,
DOI 10.1007/978-3-319-25604-7_2

To do this, we need to move beyond the old way of doing CSR and, instead, marry investments and business innovation with impact. To provide the context for where the supertankers of the global economy could be headed next, let us first look at established forms of CSR. The continuum of approaches to CSR often includes an initial focus on compliance, with an adherence to particular legal requirements. Such compliance actions are usually costly. They are not integrated into the core activities of a company. Firms have therefore typically limited activities to doing only the strict minimum required by law to reduce these costs as much as possible.

Closely related is corporate philanthropy, whereby companies "give back" a small proportion of their profit to society, sometimes through corporate foundations. A more business-minded approach is strategic corporate philanthropy, whereby firms select charitable activities aligned with business objectives so that they can add value to causes in addition to cash, albeit generally being at a net cost to the firms. Here too the scope of CSR is bound by internal limits on the amount of additional costs deemed affordable and the intrinsic volatility of profits over the business cycle.

Given the secular trend to rising transparency and lower cost of information, paired with an increasing public focus on the long-term effects of consumption and production practices, these forms of CSR no longer work as well in producing consumer and regulator goodwill, nor in NGO acquiescence. There are good reasons for this: the combination of criticisms that these more image-focused CSR approaches are merely "green-washing" strategies; the associated costs of CSR programs; and regulatory as well as activist investor requirements. Together, they have led managers to incrementally instill environmental, social, and governance (ESG) principles within corporate value creation systems. Yet this "value-based CSR" focus still misses major opportunities to drive both disruptive business innovation and value creation for stakeholders. Strategic CSR, which addresses new or untapped market needs, also remains embedded in the logic of corporate responsibility. There is a full and frequently confusing spectrum of CSR practices (Fig. 2.1).

Firms do have deep responsibilities towards the people affected by their operations. Thus there will always be a role for CSR. But the exciting new idea is to shift from the approach of solely responsibility-driven CSR to a mindset that also recognizes corporate social opportunities where there is a business benefit to doing things differently, which better captures self-interest.

True transcendence of CSR practices requires fusion of sustainable value creation and core business. This comes when external factors like consumer demand begin to have more influence on sustainability practices, product, and service design, and the resulting policies and practices start to have a positive financial return. This requires an investment and opportunity mindset. Firms that are more sensitive to the environment in which they operate have realized that socially and environmentally friendly practices are key drivers for sustainable growth and profitability.

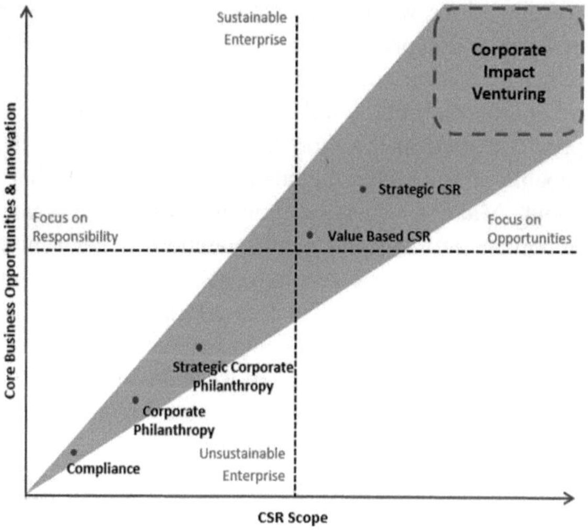

Fig. 2.1 CSR: From compliance to CIV (*source* author) [2, p. 14]

Addressing the limitations of current practice by linking what has become known as "shared value" to business innovation and impact investing is a promising pathway. "Shared value," as described by Michael Porter and Mark Kramer, is the idea that one can create economic value in a way that also creates value for society by addressing its needs and challenges [3]. Many traditional family businesses have been operating in the logic of creating both value for themselves and for their stakeholders for centuries. Yet, conceptualizing it is an important step forward. What if we build on "shared value" but shift the focus from prioritizing internal firm-based activities to (1) analyzing external trends as a source of new value creation, and (2) use an investment logic to get to disruptive new products and services to make headway with innovations that can become new business segments? [4].

CIV builds on new sources of value-added partnerships and development clusters through new products and services, value chain redesign, and production clusters that systematically produce shareholder and stakeholder value. This leads to engagement in activities such as sustainable supply chains, engagement with social entrepreneurs to generate disruptive value creation opportunities at the grassroots level, or investing in new sustainable business lines.

2.2 Play at a Higher Standard, but Compete

Some barriers stand in the way of a complete transition to a sustainable value creation strategy that fully integrates CIV to drive innovation. Businesses are highly constrained in the pursuit of corporate responsibility. The standard paradigm for business is still to be culturally and legally "hardwired" to pursue (primarily) a single bottom line of profit maximization, to be complemented by CSR considerations when convenient. This means that products and services, pricing, and production processes are still chiefly determined by their role in creating profits, rather than other societal or environmental outcomes.

Today, the notion still persists that ESG activities involve a net cost to enterprises, at least in the short term, with only a few leading companies having been capable of embracing the business opportunities that an investment logic can bring about. Make no mistake though: sustainable value creation is becoming more important and starting to get attention at the senior leadership level because of changing framework requirements, such as evolving consumer needs and total resource productivity. The four megatrends described earlier are creating opportunities also for companies in advanced economies. These trends can provide an entry point to product and service innovation that ensures long-term competitiveness in high-cost locations and enables expansion into new and adjacent markets.

Take the example of companies located in German-speaking countries. There is pressure to adapt, but there are also assets that can help. From an investor's point of view, the skills of the workforce and the social climate are part of the key competitive advantages of a country such as Germany. The situation is similar in Austria and Switzerland. Generally high ethical, environmental, and social standards are part of the social contract that protects social peace and high productivity in the region. However, this also means extra costs for companies. Some foreign companies that do not operate according to similar standards, and in industries where margins are low, have a significant short-term cost advantage. This raises the fundamental question of how to succeed in the face of international competition, while also maintaining high standards.

The relentless pressure on costs in advanced economy industries underpins that to date the magic formula how to achieve this has not yet been found. Amid the tension between short-term gains and long-term success, "soft" success factors are important and can be included in the representation of the value and performance of a company. Certain companies already carry out lifecycle assessments and prepare social balance sheets to measure social capital, which has been a requirement in France since 1977 [5]. A stable social climate confers direct advantages in international competition, and anchors long-term sustainability in the business. Lack of social peace entails considerable costs [6].

Another competitiveness factor is education. New ideas must first emerge before one can act upon them. The pride of German-speaking Europe is its dual education system [7].[2] In Germany, 60 % of all young people receive their vocational training in the dual system. Businesses and vocational schools are partners, training professionals from industry, crafts and professions of tomorrow in about 350 occupations. Anyone who has earned a diploma can easily compete with graduates in other countries. "The combination of theoretical courses and practical application in

[2]For example, the U.S. is testing the proven dual education system to reduce the shortage of skilled workers: German automobile manufacturer Volkswagen has invested USD 40 million in a new Training Academy in Chattanooga, Tennessee. The German Ministry of Economics certifies graduates of the program.

operations has made professional training into a model of success of the German education system," said the former German Education Minister Schavan [8]. Some EU countries are trying out the practical German education system in the fight against high youth unemployment. At Seat, a car manufacturer based in Spain, apprentices are learning their craft in this new way [9].

Businesses and corporate foundations are also involved directly in education as part of their corporate responsibility activities. The German Haniel Group, for example, invests heavily in promoting the field of education (e.g. schools and scholarships), the training and education of employees, and provides voluntary social benefits for employees [10]. In other words, Haniel is actively engaged in creating exactly the credentialed environment that investors value in Germany. In times of rapid change, the challenge is to translate high levels of education and technical capability into propensity for entrepreneurial value creation, rather than status quo bias.

A quality operating environment provides an excellent context for the execution of fresh entrepreneurial ideas that distil value from the megatrends discussed earlier. To be able to compete in the long term, empowering entrepreneurs who come up with new value propositions is nevertheless critical. There is a lot to learn from the buzzing entrepreneurial energy in emerging markets. There, social and environmental issues are so much "in people's face" that it seems natural to come up with business models that create both financial and social value—imagine the powerful product and service innovations that could emerge on a broad front from corporations adopting an investment and opportunity mindset to create sustainable value.

2.3 Translate Opportunity into Business Transformation

To illustrate the business transformation potential of CIV in the retail industry, consider the case of IKEA, the world's largest furniture retailer, and a pioneer of what we can term "fast furniture."

In its French business—the third largest market after Germany and the US—IKEA was facing massive reputational issues in 2013. There had been public outrage that the firm had been spying on employees and customers—this in a country that considered privacy extremely important [11]. According to The New York Times, IKEA's "investigations were conducted for various reasons, including the vetting of job applicants, efforts to build cases against employees accused of wrongdoing, and even attempts to undermine the arguments of consumers bringing complaints against the company. The going rate charged by the private investigators was 80 to 180 euros, or $110 to $247, per inquiry, court documents show. Between 2002 and 2012, the finance department of Ikea France approved more than €475,000 in invoices from investigators" [12].

Compared to that Big Brother type episode, IKEA has adopted a very different strategy to engage with its customers and employees. By 2020, IKEA plans to operate 500 stores, employ 200,000 people, welcome 1.5 billion shoppers annually and generate EUR 45-50 billion in turnover (up from 27 billion in 2012). Growing consumer demand, rising raw material prices, and ecological footprint considerations resulting from growth and changing consumer expectations are collectively motivating the company to update its business model [13].

IKEA aims to transform its business in a way that is both profitable and responsible. The company hopes to become the "leader in sustainable, affordable life at home." It has committed itself to a series of corresponding goals, including to enhance sustainable living among its employees and consumers alike, introduce new technologies, inspire new thinking by partnering with local communities, and advocate for appropriate government policies [13, pp 7–8]. According to IKEA's strategy, sustainable value creation shall become a natural part of the daily work of at least 95 % of IKEA employees by August 31, 2015. IKEA is implementing personal and incentive schemes, knowledge, training and best practice sharing to drive progress towards reaching this goal.

The company is also actively seeking recognition and is striving to position itself as a firm that takes social and environmental responsibility seriously with respect to at least 95 % of its employees and suppliers, and 70 % of its consumers. Some may consider this greenwashing, but IKEA's 2020 strategy includes hard retail stretch goals: all home furnishing materials, including packaging, are to be made from renewable, recyclable or recycled material, and all cotton used in production shall become compliant with the Better Cotton Standard [14]. Environmental commitments by August 2015 include, for instance, recycling 50 % of the non-renewable materials used in home furnishing products, and reducing carbon emissions from IKEA's operations by 50 % and those of its suppliers by 20 %. Further, IKEA plans all leather and wood used in production to be fully traceable and produced in accordance with forest and animal welfare protection guidelines and standards by the end of 2017. To achieve all of this, IKEA is encouraging its suppliers to move from compliance-driven to shared value-based social and environmental performance [15].

IKEA is a resource-hungry fast furniture retailer. Yet the IKEA sustainability strategy for 2020 is remarkable in that it covers all aspects of the business and its value chains, and sets ambitious targets. In practice, this strategy would be difficult to fully achieve by only using existing resources and business innovation mechanisms. The company will have to engage in venturing on a variety of fronts to onboard the expertise needed to make the transformation happen. To illustrate just one tiny part of that strategy, consider the company's corporate venture IKEA GreenTech AB, which has invested in six green technology companies that help IKEA "go renewable" in its core activities, encompassing energy, materials, water and waste issues [16]. Among these companies is the flax fiber producer CRAiLAR, which is also an investee of adidas' venture activities mentioned below (Spotlight 5). In 2012, IKEA invested over USD 2.8 million in the installation of equipment to support and expand CRAiLAR's European production facility and to provide working capital for IKEA orders [17].

The ambition level communicated publicly by IKEA highlights the extent to which a cohort of frontrunner companies in a number of industries sees sustainable value creation as a necessity to compete. The approach aligns well with the line of argument advanced by a number of players concerning the potential of the circular economy mentioned earlier [18]. Corresponding opportunities are all around us. Four materials categories are often cited as promising candidates for testing viability. They include: [18, p 11] (1) 'Golden Oldies' or well-established, high-volume recyclates such as paper and cardboard that still suffer from quality loss. (2) 'High Potentials,' defined as materials used in high volumes that currently lack systematic reuse solutions such as polymers. (3) 'Rough Diamonds,' consisting of large-volume by-products of many manufacturing processes, such as carbon dioxide and food waste. And (4) 'Future Blockbusters,' defined as innovative materials with breakthrough potential in terms of improving materials productivity such as 3D printing or pre-designed, fully restorative usage cycles of inputs such as bio-based materials that are returned to the biosphere [18].

For innovation on this scale to succeed in practice, efficient mechanisms to identify and nurture opportunities are key. The next chapter discusses the potential of CIV and outlines how to get started building a corporate impact program that can make good on the ambition to do more than provide the musical performance on the Titanic. Scale, speed and quality are key. To achieve business transformation at this mega scale, compared to investments made so far, we can expect that impact venturing capital allocated to finance the transformation will have to be much larger than the investment already made in a firm such as CRAiLAR. If we conservatively assume that investments of at least five percent of current turnover are the minimum threshold to set off the transformation process, it follows that investments of EUR 1.35 billion, or more than 480 times the CRAiLAR investment, are needed.

References

1. United Nations, Global Compact News (2010) A new era of sustainability: UN global compact —Accenture release finding of largest CEO research study on corporate sustainability. http://www.unglobalcompact.org/news/42-06-22-2010. Accessed 5 Feb 2014
2. Martin M (2014) Driving innovation through corporate impact venturing: a primer on business transformation, p 14. Impact Economy. http://www.impacteconomy.com/en/primer3.php. Accessed 15 Jul 2015
3. Porter ME, Kramer MR (2011) Creating shared value: how to reinvent capitalism and unleash a wave of innovation and growth. In: Harvard Bus Rev (February). http://www.hks.harvard.edu/m-rcbg/fellows/N_Lovegrove_Study_Group/Session_1/Michael_Porter_Creating_Shared_Value.pdf. Accessed 27 Feb 2014
4. Denning S (2012) Is 'shared value' a new mental model for innovation? In: Forbes. http://www.forbes.com/sites/stevedenning/2012/02/16/is-shared-value-a-new-mental-model-for-innovation. Accessed 6 Mar 2014
5. Fouquet H, Viscusi G (2010) Lagarde says pension strikes cost France as much as $563 million. http://www.bloomberg.com/news/2010-10-25/french-labor-protests-cost-economy-up-to-285-million-a-day-lagarde-says.html. Accessed 27 Feb 2014

6. Emmanuel J, Bremer C (2010) French fuel strikes falter, unions wants talks. http://www.reuters.com/article/2010/10/26/us-france-pensions-idUSTRE69P34320101026. Accessed 6 Mar 2014
7. Auswärtiges Amt—German Missions in the United States (2013) Volkswagen Trainees First in the U.S. to Receive German Apprenticeship Certificate. http://m.germany.info/Vertretung/usa/en/__pr/P__Wash/2013/08/13-VW-USA.html?r. Accessed on 6 Mar 2014
8. Deutsche Bundesregierung (2012) Duale Ausbildung. http://www.bundesregierung.de/Content/DE/Infodienst/2012/11/2012-11-05-duale-ausbildung/2012-11-05-die-duale-ausbildung-ein-flaggschiff-des-deutschen-bildungssystems.html?nn=539856#group2. Accessed 6 Feb 2014
9. Deutsche Welle (2013) Deutsche Ausbildung als Exportschlager. http://www.dw.de/deutsche-ausbildung-als-exportschlager/a-16578555. Accessed 7 Feb 2014
10. Haniel (2014) Bildung als Chance. http://www.haniel.de/de/verantwortung/holding/bildungsfoerderung/bildung-als-chance. Accessed 7 Feb 2014
11. Clark N (2013) IKEA's investigations not limited to employees. In: New York Times. http://www.nytimes.com/2013/12/16/business/international/ikeas-investigations-not-limited-to-employees.html. Accessed 15 Jul 2015
12. Clark N (2013) Revelations that IKEA spied on its employees stir outrage in France. In: New York Times. http://www.nytimes.com/2013/12/16/business/international/ikea-employee-spying-case-casts-spotlight-on-privacy-issues-in-france.html. Accessed 15 Jul 2015
13. IKEA (2013) People and planet positive: group sustainability strategy for 2020. http://www.ikea.com/ms/en_US/pdf/reports-downloads/peopleandplanetpositive.pdf. Accessed 27 Feb 2014
14. The Better Cotton Initiative (2014) The Better Cotton Initiative's register. http://bettercotton.org/about-better-cotton/better-cotton-standard-system/. Accessed 28 Feb 2014
15. IKEA (2013) People and planet positive: group sustainability strategy for 2020. http://www.ikea.com/ms/en_US/pdf/reports-downloads/peopleandplanetpositive.pdf. Accessed 27 Feb 2014
16. IKEA GreenTech AB (2014) Focus area. http://ikea.greentechab.com/focus-areas. Accessed 27 Feb 2014
17. CRAiLAR (2013) CRAiLAR and Ikea complete general supply agreement and CA$3.2 million of financing. http://www.crailar.com/company/press/crailar-and-ikea-complete-general-supply-agreement-and-ca32-million-financing. Accessed 27 Feb 2014
18. Ellen MacArthur Foundation (2013) Towards the circular economy, 1: economic and business rationale for an accelerated transition. http://www.ellenmacarthurfoundation.org/business/reports. Accessed 1 Mar 2014

3 New Pathways to Results: Corporate Impact Venturing

3.1 Fuse Economic and Socio-Environmental Performance

There is increasing visibility of the megatrends that are shifting the very foundation of our global system. The necessary recalibrations within firms to keep competing, in a dynamic interplay of simultaneous challenges and opportunities, needs to be sorted out. Yet the emerging Impact Economy is not merely a functional spectrum of opportunity, clearly segmented by instruments, actors or even issues. Rather, the sustainable capitalist future combines innovation, ideas, technology, and the difficult-to-classify. Their boundaries are inherently dynamic. They often defy conventional framing. We need to act creatively, identify innovations and translate them into business models, products, and services. Otherwise we will end up with the unfortunate conclusion, "our future—nay." To succeed, we need to be aspirational, and strong on process.

The response to these developments, and the financial opportunity contained within them, will be partially specific to the company and the industry. Successful solutions have a common denominator though: they are cross-sector by nature, require the coordination of many different stakeholders and organizations—and early success along the way will be difficult to evaluate [1].

The decisive step forward is CIV. By looking at external factors for business opportunities and deploying the mechanics of corporate venturing, a company can create disruptive business innovations that result in positive social and financial returns. This requires identifying value creation opportunities, as well as a process to identify what is sustainable, and what is not. When exploring new business territories, this distinction may not be so obvious at the outset.

Pragmatism is important. In a company, much of the business and industry insight required to identify these opportunities comes naturally from everyday business and engagement with customers and suppliers—evident in the insights gathered from an annual key suppliers meeting or interactions with especially successful clients. In many companies silent majorities of engaged employees are controlled by key

M. Martin, *Building the Impact Economy*,
DOI 10.1007/978-3-319-25604-7_3

position holders with status quo bias. They stifle innovation. The truly game-changing ideas often emanate outside corporate boundaries. They come from former employees or newcomers, who are closer to the new trends or have slightly different vantage points when it comes to value creation. It is therefore important to design a process that can produce groundbreaking opportunities in a systematic fashion, and to accept that breakthroughs rarely are 'committee material.'

Through the more detailed *how-to* discussion of corporate venturing and the process outlined in the next sections, developing a company's own sustainable value creation plan becomes possible. Such a plan must effectively leverage key groups of stakeholders and other external factors. CIV is a process for responding to constraints and opportunities, as described with respect to Patagonia and adidas below (Spotlight 5).

Spotlight 5: Patagonia—Enlist Venture Finance to Step It Up

If a company wants to build and defend a franchise where sustainable products are part of its unique selling proposition, CIV can be a great way to access new business ideas that have strategic value. With several decades of work dedicated to building a sustainable business, Patagonia, an outdoor apparel company with around USD 600 million revenues in 2013, is considered a leader in terms of sustainable value creation [2]. Patagonia has undertaken many steps over the years. These include: [3] (1) donating 1 % of the company's sales to grassroots environmental organizations. (2) Co-founding the Sustainable Apparel Coalition (SAC)—a coalition of companies producing more than a third of the clothing and footwear on the planet and committed to improving the industry's social and environmental performance through sustainability assessment tools, discussed in Part II [4]. Furthermore (3) hardwiring a commitment to social and environmental value creation next to financial performance by becoming a B Corp ("B Corp" is a US certification that incorporates a commitment to people, planet, and profits into the firm's legal charter).

The next logical step was to use the power of venture finance for sustainable business innovation. In 2013, Patagonia launched "$20 Million & Change", an internal venture fund that invests between USD 500,000 to USD 5 million per transaction into start-ups focused on clothing, food, water, energy and waste. Being "like-minded" and "environmentally responsible" was also an important part of the investment criteria [5]. Coinciding with the launch of the fund, Patagonia reorganized and established a new holding company called Patagonia Works as parent to the group's different operating businesses [6]. Launching an investment fund was—in the words of Patagonia founder Yvon Chouinard—"a logical step to reach out beyond the framework of the apparel and outdoor industries (...) to do business more responsibly" [7]. Patagonia Works currently accommodates five new lines, including Patagonia Provisions, which offers "salmon harvested only from abundant, sustainable fisheries"; Patagonia Media, which supports production of awareness raising media materials; and Fletcher Chouinard Designs, which produces surfboards from a minimum use of toxic and nonrenewable materials with the goal of "causing no unnecessary harm," among others [8–10].

Another example of an in-house investment fund is the approach of adidas, a clothing and footwear company with EUR 14.5 billion in annual turnover (2014) [11]. In an attempt to access new business opportunities, adidas created Hydra Ventures in 2011. The Netherlands-based vehicle focuses on seed investments in apparel brands and sports related areas such as sports technology. While the investment criteria are mainly financial giving consideration to the environmental sustainability and social performance record of investees creates value in practice [12]. For example, in 2013, Hydra Ventures invested nearly USD 1.8 million in CRAiLAR Technologies Inc., the company mentioned earlier that provides sustainable, environmentally friendly fibers and fabrics [13]. CRAiLAR's fully owned subsidiaries develop BastFiber technology for use in textiles, cellulose pulp, paper, and composites, and provides environmentally sustainable hemp, bamboo, organic cotton and soy blended apparel [14]. Prior to Hydra Ventures, adidas had already established the greenENERGY Fund to facilitate profitable and value-creating investments that respond to the industry problem of energy efficiency [15]. Its scale is tiny, but with the greenENERGY Fund, adidas aims to pursue the goal of (1) accelerating carbon reduction in its global properties, and to also (2) deliver a 20 % plus return on capital. The approach highlights the potential of venturing when the parent company is also the first client [16]. The fund's first seven projects results' are modest, but interesting: they cut carbon by 1401 metric tons of CO_2 (comparable to 256 fewer cars on the road each year) [17]. The logical next step is to deploy this kind of thinking at scale.

3.2 Graduate from VC to CIV

Large corporations often find it difficult to bring innovations to the market, even if their deep industry expertise enables them to perceive many trends and business ideas first. In response, classical corporate venturing emerged in the 1960s. Corporations were looking for a way to better execute on innovation rather than see their employees resign only to then turn fresh business ideas into blockbuster successes outside the corporation [18].

The underlying insight is straightforward. VC has been a driver of product and service innovation since the conclusion of World War II. When larger companies invest in start-ups or spin-offs, they engage in corporate venturing. They have three options: (1) setting up an internal corporate venturing group that invests off balance sheet, (2) creating a dedicated external corporate venturing fund, or (3) becoming a limited partner (LP) in one or several venture funds that follow investment strategies, which are relevant to the corporation. As in the case of any investment, it is important to be clear about investment objectives, instruments and processes as well as investment sourcing and geographical focus.

Private venture capitalists have often derided corporate venturing, criticizing the intrusion of corporate bureaucracy in the investment process and the lack of clear priorities. Corporate venturing typically seeks to balance the parent company's strategic and financial considerations. The relative weights can change in the process, depending on who is running the show at the senior management level.

Corporate venturing cycles have historically closely tracked the boom and bust of the independent VC sector. Moreover, corporations have a tendency to overbuild investment capacity as the business cycle unfolds without carefully thinking through the implications, often having to dismantle programs before they have had enough time to achieve successful investments [18]. As suggested in a 2014 Harvard Business Review article, the median life span of corporate venturing programs has traditionally been as low as one year [19].

Notwithstanding, corporate venturing is an important innovation mechanism, growing in significance when firms need to reinvent their business models. In the US alone, venture capitalists invested USD 48.3 billion in 4356 deals in 2014, which is an increase of 61 % in USD and a 4 % increase in deals over the prior year, marking the highest level in over a decade [20]. The majority of corporate venturing deals took place in the US, typically in IT or other sectors driven by IT innovation. European countries—with the UK in the pole position—and Asian countries are catching up. Although Europe accounted for only 15 % of global VC activity, 2013 was a watershed year for the European VC industry. The number of investment rounds was up by more than 5 % on the prior year and the value of investments rose by 19 % to USD 7.4 billion [21]. In 2013, 72 % of new corporate venture companies had a non-US parent company, with 38 % stemming from Europe and 28 % from Asia [22].

As part of its 2014 Innovation Plan, the European Commission (EC) was considering innovative co-investment schemes to further enhance corporate venturing in Europe [23]. In an interesting reversal of the trend in the US and European markets, Chinese VCs started looking toward earlier stage investments with more competition between a range of investors interested in middle and later stage companies thus driving up valuations [24, p 20]. With more than 180 new corporate venture funds raised between 2010 and 2012, and around 60 % of the 1090 corporate venture firms located outside the US in 2012, corporate venturing has become truly global [24].

Synergies that a corporation can develop with its investments are the main driver of value creation in corporate venturing. Compared to building R&D capacity in-house, Corporate Venture Capital (CVC) can be much cheaper and faster—provided the parent company manages to benefit from the know-how it invests in. Positive financial returns are then a plus, but not the only success metric of the business case for corporate venturing. For example, in the classical corporate venturing space, Intel Capital has done very well by financing start-ups that build an ecosystem for the flagship products of the semi-conductors specialist.

The megatrends discussed earlier add the dimension of social and environmental performance to value creation, but an effective approach to innovation on a broad front is needed in order to master this equation. A logical step is to apply the tools of corporate VC to the world of impact—and to engage in CIV. CIV is situated at a mid-point between classical CSR. CSR is mainly motivated by the company's responsibilities vis-à-vis its stakeholders. Traditional CVC seeks to secure access to the business innovations the company needs to be successful in tomorrow's economy. In a world where social and economic value creation are converging, venturing makes sense to drive both.

3.3 Move Impact Ambitions from Niche to Scale

Any corporate impact investment strategy should leverage the key drivers of value creation in corporate venturing: the synergies that a corporation can develop with potential investments. Most start-ups struggle in the go-to market phase of their development as qualified sales specialists are expensive, established customers are wary to adopt products and services from unproven and young companies, and sales cycles can be lengthy. Hence, a partnership with a large corporation can deliver immediate and significant benefits for any type of start-up, whether impact-focused or not. The partnering corporation can be a client, a beta-tester or a first customer. It can also open doors for the start-up with other corporations, reducing the length of the sales cycle. Often enough, the start-up can tap additional resources such as advice and sometimes executives from the corporation. The company is subsequently a potential buyer for the start-up. Companies can harness external factors, trends, or ideas for sustainable business growth through impact venturing. They invest in new models for sustainable value creation. These can then be tested and later transferred into the core business once they are perfected—or cloned.

Talents are also voting with their feet. Many seek to join firms that provide a space for combining impact and entrepreneurship. This has led to the emergence of synthetic social entrepreneurs or "social intrapreneurs"—employees who act in entrepreneurial ways to update some aspect of the value proposition of the company where they work. This is typically around a topic linked to one of the sustainability megatrends discussed earlier. They think of themselves as "intrapreneurs" (in analogy to entrepreneurs), rather than just employees [25, 26].

However, there are a few questions to consider when contemplating engaging in CIV. One is how to deal with financial profit. When considering the relationship between the core business and impact, a CEO has the fundamental choice to (1) interpret "impact" in a zero-profit sense and to then reinvest any proceeds of the corporate impact venture. Or, (2) to interpret it in a for-profit sense and focus on raising social as well as environmental performance through innovation that impacts some core aspect of the business, but allows for earning the cost of capital. Another option is (3) to combine both strategies, where the majority of profits are reinvested for business development (such as with the zero-profit approach), but still acknowledging the need to provide economic sustainability while at the same time transferring the social benefits to the core business.

To get a sense of the state of play in corporations' involvement in impact investing, consider the cases of Danone and Uniqlo, both working in partnership with Grameen in Bangladesh. Following Nobel Peace Prize winner Muhammad Yunus' approach to social businesses, global food company Danone supports, both financially and operationally, selected projects through the Fonds commun de placement à risques (FCPR), a French VC fund called "danone.communities" [27]. The selected projects, according to Yunus' social business definition, should: (1) have a social objective, (2) be a non-loss company and (3) be a no-dividend company in which "the profits are reinvested in the business to increase social impact" [28]. It is based on this zero-profit approach that global food company Danone and global retailer Uniqlo have approached their respective venture partnerships in Bangladesh with social sector organization Grameen [29, 30].

In the case of Danone, the venture uses a combined strategy that allows to directly transfer the social benefits to the core business by re-investing the profits in "expanding and running the business" with criteria aimed at improving livelihoods and the environment. Compare this to Uniqlo, which reinvests all its profits into social business initiatives [31, 32]. Grameen Danone produces affordable and fortified yoghurts whereas Uniqlo produces affordable clothes. This has the potential to create greater prosperity and serve poor consumers, helping Bangladesh achieve significant development.

To find out the relevance of such activities, the question is their ultimate scale. In its National Sustainable Development Strategy for 2011–2021, the government aims for Bangladesh to become a middle-income country by 2021 via sustainable development [33]. Garment making is Bangladesh's largest export earner and engine of growth, and a topic we will examine in detail in Part II. In 2005/06, the food processing industry contributed over 13 % to total industrial value in Bangladesh, and ranked third after the Ready-Made Garment (RMG) and gas industries [34]. The food-processing industry is more than ten times the size of the apparel sector and thus an interesting candidate for innovation: Bangladesh still faces significant social and environmental challenges. These include extreme poverty and food insecurity, with one of the highest malnutrition rates in the world [35].

Given its purpose to tackle malnutrition, Grameen Danone Foods Ltd. was developed with the stated objective to serve the rural poor. Produced in a local and environmentally friendly factory where solar energy is used to heat cleaning water, Grameen Danone yoghurts were significantly cheaper than competitor products. Local women sold them to generate complementary income. After two years of operation at the BoP, the joint venture had not yet managed to achieve breakeven, whereas Danone made a profit of more than one billion dollars in 2008 [36, 37]. In response, the social business closed the direct distribution channel via local women. Instead, it started marketing the yoghurt through established retail channels located in urbanized areas, with some success [38]. Another project, La Laiterie du Berger in Senegal—a manufacturer and distributor of dairy products—is yet another example where both Danone and The Grameen Foundation are currently investing [39, 40].

In contrast to Danone's focus on nutrition, Uniqlo's joint venture Grameen Uniqlo focuses on tackling another dimension of poverty: producing clothing for the poor. According to the company, all product related processes are completed in Bangladesh, including procurement of materials and sales. Selling items on a door-to-door basis helps women earn money. Profit generated in the seven Grameen Uniqlo shops is reinvested in additional social business initiatives [41]. According to Yukihiro Nitta, the CEO of Uniqlo Social Business Bangladesh Ltd, Grameen Uniqlo is Uniqlo's "commitment to eliminate various kinds of social problems" [42].

No significant financial returns can be expected in both cases, and mainly in the case of Uniqlo with the pure zero-profit approach. However, the return on brand image has been very substantial, especially in the case of Grameen Danone. The venture received the Vision Award 2009 for social innovation [43]. Bloomberg BusinessWeek listed it among 25 products that have the potential to change the world [44]. Besides conventional marketing techniques, Grameen Danone successfully addressed the social aspect while marketing yoghurt to wealthier consumers, raising its sales up to 80 % in urban areas in April 2010 [45].

The approach has the advantage of being able to dispel potential criticism and sticky topics such as "profiting from the poor." However, the zero-profit approach in the case of Uniqlo and the combined strategy implemented by Danone, raise another significant question: if an impact venture is designed *not* to cover the cost of capital, how will its value proposition ever attract additional capital, beyond subsidies? Can it be reasonably expected to scale to a point where it has a chance to transform the business of the parent company or the industry, especially if it serves the BoP customer segment on low margins?

If we are serious about converging on an Impact Economy on a major scale, we will need to find ways to finance large-scale opportunities and compensate capital adequately for the risks we invite it to take on. Otherwise, we will remain confined to a niche.

3.4 Build a Program for Business Transformation

Supertankers cannot be turned around like speedboats. CIV is an interesting way to access innovations. Operationalizing it is key however. A simple formula to develop an appropriate response to the trends unfolding does not (yet?) exist. However, there are some straightforward steps to go through to develop a corporate impact program that can drive profits and social and environmental performance. Being disruptive is a necessary, but not a sufficient, condition for success. Three other factors are key: getting the right people onboard that have an entrepreneurial mindset, patient capital, and risk tolerance. Models need testing to gather evidence of significant market potential and financial return. Under a commercial return lens, good ideas are often not given sufficient time to be tested. Senior leadership and other select internal players need to be behind any ambitious activity that treads

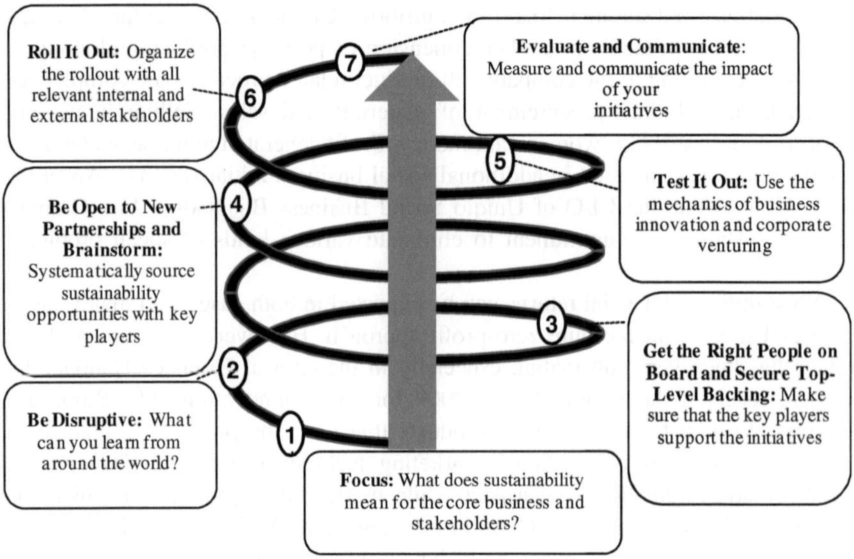

Fig. 3.1 Key steps toward business transformation (*source* author) [46, p. 23]

new ground. Without their backing and understanding of the strategic importance of innovation and sustainable value creation (and resulting insights), the initiative cannot succeed. Here are the steps (Fig. 3.1).

Focus. Encourage exploration of new approaches in areas aligned with the core business. This is the best way to harness market knowledge, internal skills and convening power. As you zoom in, research the issue and industry to understand the full context of the situation. Besides classical industry analysis, this also includes understanding how the broader concepts of sustainability (shared value theory, ethnic and gender dynamics, etc.) play out in the communities and markets you will work in. Think of this as a new solution, rather than as a new "project" that will have everyone wondering when it will finally be over.

Be Disruptive. Look globally to surface similar business models in the same or different regions in order to tease out common challenges and potential lessons that can be applied to your specific context, such as barriers to growth or issues in supply chains. The key to success is to test and scale new business models, often with new partners. While naturally looking for financial viability, it is important to recognize and accept that the new net double-digit internal rate of return (IRR) models could be the exception, not the norm.

Get the Right People on Board and Secure Top-Level Backing. Being disruptive is a necessary, but not a sufficient, condition for success. Three other factors are key: getting the right people onboard that have an entrepreneurial mindset, patient capital, and risk tolerance. Models need testing to gather evidence of significant market potential and financial return. Under a commercial return lens, good ideas are often not given sufficient time to be tested. Senior leadership and other select internal players need to be behind any ambitious activity that treads new ground. Without their backing and understanding of the strategic importance of innovation and sustainable value creation (and resulting insights), the initiative cannot succeed.

Be Open to New Partnerships and Brainstorm. Successful models that are genuinely new are typically developed when different skill sets come together, for example via corporate partnerships with NGOs or social businesses. This requires recognizing such organizations as partners with an aligned vision and objectives, not merely seeing them as contractors. Openness, flexibility and shared goals are essential. This is easier said than done of course, and entails blending "business Deoxyribonucleic Acid" (DNA) with "development DNA." To operationalize insight generation and model development, run workshops with the key players in order to facilitate the formation of sustainable value creation opportunities. You need a fertile and open environment to have successful brainstorming sessions, with a broad range of options to narrow down to relevant solutions.

Test It Out. After the facilitation process, test what you have sourced or developed to ensure success. Use the mechanics of business innovation and corporate venturing to your advantage. The most sustainable business innovations often emerge from merging internal insights with disruptive spin-offs or start-ups. As you go about testing, do not forget that new models require patience and flexibility during the test, incubation and growth phases.

Roll It Out. After honing in on one (or a few) possible initiative(s), organize the rollout strategy and make sure key internal decision makers assume ownership. Bring in all relevant external stakeholders at this early stage to develop a realistic implementation plan with cost controls. Do not forget that your solutions are likely to encounter a suite of market and policy barriers that will need to be tackled equally, such as financing, market entry and regulatory barriers. We often fall victim to the illusion that nurturing pioneering solutions alone is enough. However, even the best solutions typically need a hand in unblocking other market barriers in order to succeed at scale.

Evaluate and Communicate. Finally, measure the impact of your initiatives or programs to maximize the efficiency and effectiveness of the activities, and optimize the ability to communicate progress achieved with key stakeholders. At no point in time should you underestimate the challenges you will encounter. Be patient, flexible and entrepreneurial.

Fresh ways to solve very large problems are urgently needed. Big business can help pull several key levers to help build economies and societies that are vastly more viable over the long haul. Yet, one needs to get past shareholders who are hell-bent on relatively short-term returns.

This book offers solutions how to succeed. The new paradigm that can take shape faces many challenges. It also offers untold promise. There is ultimately no alternative. Many people in leadership positions are aware of this. Paul Polman, CEO of Unilever, famously commented "Our version of capitalism has reached its sell-by date. Never has the opportunity for business to help shape a more equitable future been so great" [47].

Corporate immune systems, key position holders and many shareholders may be reticent to change. Part I nevertheless explored a pathway to extend capitalism's lease through innovation and investment, responding to sustainability megatrends. In Part II, let us now be very practical. We will look at the example of the textile and garment industry: how can we align the supply chains of this emblematic industry with our expectations for social and environmental performance in the first half of the twenty-first century, while remaining competitive? Turn the page to find out.

References

1. Hallie P, Beer T (2012) Evaluating social innovation. FSG and the Center for Evaluation Innovation. http://www.fsg.org/tabid/191/ArticleId/708/Default.aspx?srpush=true. Accessed 27 Feb 2014
2. Drake B (2014) How Patagonia's new CEO is increasing profits while trying to save the world. In: Fast Company. http://www.fastcompany.com/3026713/lessons-learned/how-patagonias-new-ceo-is-increasing-profits-while-trying-to-save-the-world. Accessed 6 Mar 2014
3. Chouinard Y et al (2011) The sustainable economy. In: Harvard Bus Rev. http://hbr.org/2011/10/the-sustainable-economy/ar/1. Accessed 28 Feb 2014
4. Sustainable Apparel Coalition (2014) The Apparel Coalition register. http://www.apparelcoalition.org/higgindex/. Accessed 28 Feb 2014
5. Patagonia Works (2014) Patagonia launches '$20 million and change' and Patagonia works—a holding company for the environment. http://www.patagonia.com/pdf/en_US/PATAGONIA_LAUNCHES_20M_CHANGE_AND_PATAGONIA_WORKS.pdf. Accessed 27 Feb 2014
6. Winter C (2013) Patagonia's latest product: a venture fund. In: Bloomberg Business Week. http://www.businessweek.com/articles/2013-05-06/patagonias-latest-product-an-in-house-venture-fund. Accessed 27 Feb 2014
7. Patagonia Works (2013) Patagonia launches '$20 million and change' and Patagonia works—a holding company for the environment. http://www.patagonia.com/pdf/en_US/PATAGONIA_LAUNCHES_20M_CHANGE_AND_PATAGONIA_WORKS.pdf. Accessed 27 Feb 2014
8. Patagonia Works (2014) The Patagonia works register. http://www.patagoniaworks.com/docs/FAQ.pdf. Accessed 27 Feb 2014
9. Fletcher Chouinard Designs (2014) FCD technology, surf technology. http://www.fcdsurfboards.com/about-fcd. Accessed 13 Mar 2014
10. Patagonia Provisions (2014) This is salmon. http://www.patagoniaprovisions.com/pages/salmon. Accessed 13 Mar 2014
11. adidas Group (2014) Make a difference. In: adidas Group annual report 2014. http://www.adidas-group.com/media/filer_public/2b/2f/2b2fd619-5444-4ee8-9c07-baa878d658c4/2014_gb_en.pdf. Accessed 22 May 2015
12. Hydra Ventures (2014) About Hydra Ventures: overview. http://www.hydra-ventures.com/about-us/index.html. Accessed 27 Feb 2014
13. CRAiLAR (2013) CRAiLAR receives CAD$2.0 million in private placement from Hydra Ventures B.V. http://www.crailar.com/company/press/crailar-receives-cad20-million-private-placement-hydra-ventures-bv. Accessed 27 Feb 2014
14. New York Times (2008) CRAiLAR Technologies Inc.: company information. http://topics.nytimes.com/top/news/business/companies/naturally-advanced-technologies-inc. Accessed 27 Feb 2014

15. adidas Group (2013) The adidas Group greenENERGY Fund: key facts. http://blog.adidas-group.com/wp-content/uploads/2013/04/adidas-Group-Green-energy-Fund-Fact-sheet.pdf. Accessed 27 Feb 2014
16. Turnbull E (2013) How the adidas Group prioritizes investments in energy efficiency. In: 2 degrees community. https://www.2degreesnetwork.com/groups/2degrees-community/resources/how-adidas-group-prioritizes-investments-energy-efficiency/. Accessed 27 Feb 2014
17. Turnbull E (2013) adidas Group scores big with sustainability venture capital fund, GreenBiz.com. http://www.greenbiz.com/blog/2013/05/15/adidas-group-scores-big-sustainability-venture-capital-fund?page=0%2C0. Accessed 27 Feb 2014
18. Gompers PA (2002) Corporations and the financing of innovation: the corporate venturing experience. Federal Reserve Bank of Atlanta. https://www.frbatlanta.org/filelegacydocs/policy_gompers.pdf?origin=publication. Accessed 25 Feb 2014
19. Lerner J (2013) Corporate venturing in: Harvard Bus Rev. http://hbr.org/2013/10/corporate-venturing/ar/1. Accessed 25 Feb 2014
20. National Venture Capital Association and PwC (2015) Press release: annual venture capital investment tops $48 billion in 2014, reaching highest level in over a decade, according to the MoneyTree Report. http://nvca.org/pressreleases/annual-venture-capital-investment-tops-48-billion-2014-reaching-highest-level-decade-according-moneytree-report. Accessed 22 May 2015
21. Ernst and Young (2014) Adapting and evolving: global venture capital insights and trends 2014. http://www.ey.com/Publication/vwLUAssets/Global_venture_capital_insights_and_trends_2014/$FILE/EY_Global_VC_insights_and_trends_report_2014.pdf. Accessed 26 May 2015
22. Global Corporate Venturing (2014) In: 2013: full year data analysis. http://www.globalcorporateventuring.com/article.php/7554/data-analysis-beyond-the-big-3#sthash.7NHThE96.dpuf. Accessed 25 Feb 2014
23. Mawson J (2012) Corporate venturing in the UK. In: RSA. http://www.thersa.org/__data/assets/pdf_file/0008/636830/Corporate-Venturing-report.pdf. Accessed 25 Feb 2014
24. Global Corporate Venturing (2013) CVII infographic: corporate venture. http://www.globalcorporateventuring.com/pages/cvii-infographic.html. Accessed 25 Feb 2014
25. Martin M (2011) Four revolutions in global philanthropy. Impact Economy. http://www.impacteconomy.com/en/wp1.php. Accessed 25 Feb 2014
26. Agoada J (2014) 2014's Most valuable employee: the social intrapreneur. In: Forbes http://www.forbes.com/sites/ashoka/2014/01/24/2014s-most-valuable-employee-the-social-intrapreneur/. Accessed 7 Mar 2014
27. Danone Communities (2014) How does it work? http://www.danonecommunities.com/en/content/danonecommunities-how-does-it-work. Accessed 12 Mar 2014
28. Danone Communities (2008) Muhammad Yunus and social business. http://www.danonecommunities.com/en/content/muhammad-yunus-and-social-business. Accessed 12 Mar 2014
29. Danone Communities (2014) Grameen Danone Foods Ltd. http://www.danonecommunities.com/en/project/grameen-danone-foods-ltd. Accessed 25 Feb 2014
30. Grameen Uniqlo (2014) About Grameen Uniqlo. http://www.grameenuniqlo.com/en/. Accessed 25 Feb 2014
31. Danone Communities (2014) Grameen Danone Foods Ltd: The project's history. http://www.danonecommunities.com/en/project/grameen-danone-foods-ltd?mode=history. Accessed 6 Mar 2014
32. Grameen Uniqlo (2014) Social business framework. http://www.grameenuniqlo.com/en/about/business.html. Accessed 6 Mar 2014
33. The Bangladeshi government formulates specific strategies and implementation tasks via two five-year plans: The Sixth Five Year Plan (2011–2015) and the Seventh Five Year Plan (2016–2020). Ministry of Environment and Forests, People's Republic of Bangladesh (2012)

Rio + 20: National Report on Sustainable Development, p 10. http://sustainabledevelopment.un.org/content/documents/981bangladesh.pdf. Accessed on 25 Feb 2014

34. USDA Foreign Agricultural Service (2013) Food Processing Industries in Bangladesh. http://gain.fas.usda.gov/Recent%20GAIN%20Publications/Food%20Processing%20Industries%20in%20Bangladesh_New%20Delhi_Bangladesh_7-3-2013.pdf. Accessed 15 May 2015
35. Hubbard G (2013) Fighting malnutrition in Bangladesh with education. UNICEF. http://www.unicef.org/infobycountry/bangladesh_68757.html. Accessed 6 Mar 2014
36. Melik J (2009) Danone's yogurt strategy for Bangladesh. In: BBC News. http://news.bbc.co.uk/2/hi/8100183.stm. Accessed 28 Feb 2014
37. The Grameen Creative Lab (2014) Grameen Danone Foods Ltd. http://www.grameencreativelab.com/live-examples/grameen-danone-foods-ltd.htm. Accessed 6 Mar 2014
38. Bapat P (2011) Failure or success waiting to happen? The case of Grameen Danone. http://www.hunger-undernutrition.org/blog/2011/06/failure-or-success-waiting-to-happen-the-case-of-grameen-danone.html. Accessed 28 Feb 2014
39. Grameen Crédit Agricole (2014) Our approach: what is social business? http://www.grameen-credit-agricole.org/en/content/our-approach-0. Accessed 12 Mar 2014
40. Danone Communities (2014) La Laiterie du Berger. http://www.danonecommunities.com/en/project/la-laiterie-du-berger. Accessed 12 Mar 2014
41. Grameen Uniqlo (2014) Brand concept. http://www.grameenuniqlo.com/en/about/index.html. Accessed 28 Feb 2014
42. Sathi, MC (2013) Grameen Uniqlo. In: Dhaka Tribune. http://www.dhakatribune.com/commerce/2013/jun/16/grameen-uniqlo-launch-apparel-stores-july-5. Accessed 25 Feb 2014
43. Vision Summit (2014) In: the visit summit register. http://www.visionsummit.org/vision-award.html. Accessed 28 Feb 2014
44. Pilloton E (2009) 25 Products that might just change the world. In: Bloomberg Business Week. http://images.businessweek.com/ss/09/12/1209_25_world_changing_products/12.htm. Accessed 28 Feb 2014
45. Robinson E (2012) Can big business sell nutritious food to poor people? The case of Grameen Danone Foods in Bangladesh. In: IDS Globalization and Development. http://www.globalisationanddevelopment.com/2012/10/can-big-business-sell-nutritious-food.html. Accessed 28 Feb 2014
46. Martin M (2014) Driving Innovation through Corporate Impact Venturing: A primer on business transformation, p 23. Impact Economy. http://www.impacteconomy.com/en/primer3.php. Accessed 15 Jul 2015
47. Wall Street Journal (2012) Three perspectives on 2012. http://online.wsj.com/article/SB10001424052970204464404577112702902409714.html. Accessed 6 Mar 2014

Part II
Fashion: Upgrade Supply Chains Versus Dress to Kill

There is no beauty in the finest cloth if it makes hunger and unhappiness.

Mahatma Gandhi[1]

[1]Quote by the Mahatma Gandhi character in the movie "Gandhi" (1982). http://www.imdb.com/title/tt0083987/quotes. Accessed 17 July 2015.

Fix Textile and Garment Supply Chains

4

4.1 Value or Disaster: What's It Going to Be?

Clothing and footwear are our chosen second skin. Next to providing cover, they are essential to our view of who we are, and who we desire to be. The textile and garment industry is also the world's #2 polluting industry, after the petroleum industry [1]. How do these two things go together, and what can we do about it?

Just take the one billion sports shoes produced every year. People buy them to walk comfortably and exercise. Like millions of people, I just went running this morning. Most of the shoes contain synthetic leather made from polyurethane (PU). In its report "a red card for sportswear brands," global environmental organization Greenpeace reported on results from testing 33 items of football merchandise ahead of the 2014 FIFA World Cup in Brazil. Part of its wider "Detox" campaign asking global apparel and footwear brands to stop using harmful chemicals, Greenpeace tested boots, goalkeeper gloves and the official 'Brazuca' ball. It found a range of hazardous chemicals in the final products. These included N,N-Dimethylformamide (DMF) in all 21 boots scrutinized [2]. DMF is the leading solvent used in the production of any PU coated textiles such as sports footwear. A colorless liquid that mixes with water and most organic liquids, it is also used in artificial leather, rain and protection wear production more generally.

DMF also happens to be a substance of very high concern (SVHC) under the European chemicals regulation REACH [3]. The EU imports and manufactures between 10,000 and 100,000 tons of DMF per year; roughly 25 % are used in the textile and garment industry [3]. But of course, most textile and garment manufacturing takes place outside the EU borders. When DMF comes in contact with skin, it is classified as toxic. Under long-term occupational exposure, it can also cause damage to the liver and is toxic to reproduction. In short, DMF is unpleasant, dangerous stuff.

While Greenpeace's call to action focused on the final product, the comparatively bigger problem resides at the manufacturing stage. PU synthetics are used to produce seven billion pairs of fashion shoes, two billion bags, one billion fashion garments, and one billion pairs of sports shoes every year. 99 % of PU synthetics

M. Martin, *Building the Impact Economy*,
DOI 10.1007/978-3-319-25604-7_4

factories use DMF as the process solvent. Four billion square meters of PU synthetic create 12.3 billion l of DMF-contaminated wastewater and 30,000 tons of solid waste [4, 5]. A change in the production process holds the solution. While this may sound utterly counterintuitive, shifting from DMF to a waterborne production process could actually *save* water. For example, Bayer MaterialScience's INSQIN® waterborne PU technology allows for solvent-free manufacturing, which—next to eliminating the environmental and health hazard from DMF—, can save up to 95 % of water and up to 50 % of energy [4, 5]. For more on chemicals and textiles, see the Spotlight 6 at the end of this section. For now, let's not get ahead of the story how we could turn around the global textile and garment industry, and start at the beginning instead.

Defined as comprising textiles, garments and luxury, the global textile and garment industry reached a turnover of almost three trillion dollars in 2011 [6]. The sheer size of the industry is impressive and so too is the omnipresent role its products play in our daily lives. Apparel is a forerunner of globalization: it was one of the first industries to adopt a global dimension, incorporating developing countries into the supply chain. Considered an important catalyst for national development and industrialization, apparel has been a natural starter industry for export-oriented countries [7–9]. The industry has been a stepping-stone to development in virtually all countries since it kicked off the industrial revolution in the United Kingdom (UK) 250 years ago. The industry is the perfect illustration of the *how-to* questions we need to solve if we want to upgrade the supply chains that connect developing countries and advanced economies, and reconcile economic, environmental and social performance.

Today, the textile and garment industry is undergoing profound change. It is both reflecting and driving the transformation of the global economy and society. The industry proffers a great window on our problems in charting a more sustainable path forward. It has closely mirrored the general post-modernization of consumption and production. Products are turned around faster and faster. They become smaller in the case of some consumer goods such as electronics. Garments and footwear are often produced in supply chains spread around the globe. Accelerating product innovation cycles and offshoring have also served to highlight the longstanding and intensifying sustainability challenges in the industry. Products are frequently manufactured under labor conditions that would be illegal in advanced economies. Shrinking product use lifecycles mean that garments are consigned to waste faster.

The net of these challenges is pressure on social and environmental conditions. This state of affairs raises uncomfortable questions. Lower-skilled workers end up operating chemicals, factories, and machinery they are otherwise untrained to handle. They have to turn out garments in a breathless rhythm, even though their wages do not cover their cost of living. As a result, social and environmental problems have been growing in severity and pervasiveness in apparel producing countries. The Rana Plaza factory disaster that killed 1133 people in Dhaka, Bangladesh in April 2013 served as the most blatant example [10]. The accident led buyers to take action on safety in the textile and garment industry in Bangladesh.

They committed to a mainly European buyer-funded Accord on Fire and Building Safety in Bangladesh (Accord) on the one hand, and the Alliance for Bangladesh Worker Safety led by US buyers on the other, both focusing on the most visible problem, namely building and fire safety [11, 12].

Part II argues that we are approaching a turnaround opportunity for social and environmental impact. In fashion products, most of the value is in the brand. The order dockets found after the Rana Plaza disaster in the rubble showed that the clothes made there cost about a tenth of their retail price [13]. In more exclusive market segments, this ratio can jump to 20 times or more the price at the factory gate. A big cost driver at the manufacturing level are fabrics. In denim products, the cost of the fabric can make up for 60 % of the cost of the shipped product (free on board, FOB), and about 70 % of the cost of goods sold (COGS) [14]. Some market segments support higher factory gate prices. For example, Keff NYC, a New York-based knitting company that supplies higher-price retailers such as Abercrombie, Anthropologie and Ralph Lauren, produces in the US and markets its "Made in USA" label [15]. Its labor costs are about 40 % higher than in China. Retail prices end up 20 % higher.

For some segments, higher end prices may work, but volume strategies require a different route. Leading sportswear company Nike has been reported to produce between 70 and 76 % of its apparel and 85 % of its footwear products on lines using lean manufacturing; this delivered additional savings of USD 0.15 per unit through better labor productivity and lower waste in 2013 [16]. All in all this may sound like small change. However, in an industry shipping billions of items, small change adds up, and the environmental and social gains of better manufacturing practices are disproportionate. To graduate from a mere diagnosis of problems, the answer lies in the merger of vastly improved social and environmental performance and growth and competitiveness. Significantly, while the global textile and garment industry's problems are especially visible, the complex supply chains of other labor-intensive industries pose many similar challenges. Apparel value chains can therefore offer a blueprint for solutions that work at scale, especially if these solutions can be financed on capital markets, based on their ability to return commercial value whilst attacking some of the worst social and ecological deficiencies of their industry.

Change is upon us, so let's make it work to our long-term advantage. Since the 1990s, the advent of fast fashion in particular—referring to fashion products that move quickly from the catwalk to high street stores in order to capitalize on current fashion trends—has led to the transformation of mass consumption and production patterns of apparel in the main markets of Europe, North America and Japan [17]. Previously dusty fashion retail groups have grown in turnover and profits to a point where they now capture the fascination of the world of investments, which is looking for high returns and mega growth stories.

Once a fast fashion brand has achieved maximum consumption speed, there are three next logical growth steps. First, to move up market, capturing the consumer segments that can afford to pay more and that traditionally shop for luxury items. Second, to expand spatially and roll out fast fashion in emerging markets where new middle classes also want to express identity and success through garments.

Third, to sell products to women that were formerly geared for men, and vice versa. Pioneers of fast fashion such as H&M, Zara and Uniqlo are doing all of this. They hire high fashion designers to move their brands up market and to better understand value creation in the luxury industry [18]. China, the focus of Spotlight 7 further on, is forecast to account for 44 % of all global sales of luxury products by 2020, up from 15 % in 2011 [19]. Convergence is happening in other ways too: next to unisex clothing, women have started shopping for watches, and men have begun buying perfumes. The lines of gender demarcation are becoming blurred.

There is a fundamental challenge to this story of growth and innovation though: the current path of the industry is unsustainable. Through Rana Plaza, people have become increasingly aware of this. The largest industrial accident in the region since the Bhopal disaster in 1984 sent ripples across the world. The effects of negative media coverage at the country or regional level triggered scrutiny and industry-wide action to address fire and building safety. This led to the creation of three initiatives: the European brand-led Bangladesh Accord on Fire and Building Safety (Accord), the American brand-led Alliance for Bangladesh Worker Safety (Alliance), and Bangladesh's National Tripartite Plan of Action on Fire Safety and Structural Integrity (NTPA) [20, 21]. Two years after the disaster, they had screened over 2000 readymade garment factories in terms of their structural, electrical and fire safety (SEF), and started to implement corrective action plans (CAP). For example, by February 28, 2015, the Accord had inspected 1405 factories (of which 225 jointly with the Alliance), and approved 817 corrective action plans [22].

While a valuable step, the question remains, who will finance and implement all the upgrades needed just at the manufacturing level in a country such as Bangladesh? We need to acknowledge that (1) the industry's problems are deep and systemic and therefore need to be addressed at the root cause level. Further, as in the case of replacing DMFs with waterborne technologies mentioned earlier, (2) the solutions around the corner could be much more powerful than we typically care to imagine. Systemic action is the way forward. As the reader will see, such action is now possible.

Next to systematic, we also need to be realistic and strategic to get to a higher-level equilibrium in terms of social, environmental and economic performance. Currently, though, most actors only focus on aspects of the global textile and garment industry (e.g. living wages, child labor, health and safety, aspects of the environmental footprint). They are working on some dimension of a very large problem and limiting their focus to narrow agendas. Much less effort has been spent on understanding and calibrating actions and strategies to maximize the prospect of overall systems change.

The dominant mindset is a significant barrier that prevents apparel from becoming a truly sustainable industry where long-lasting value creation is at the core of business. Moreover, we cannot turn back the clock. Arguments that consumers should shop less are voiced every once in a while, but evidence shows that consumption will not decelerate, absent a major war occurring [23]. In the US, as well as the UK, the average consumer throws away 30 kg of clothes every year, and our wardrobes hold several times the amount of clothes that our grandparents owned [24, 25]. Many emerging middle-class consumers in developing countries similarly aim to expand their wardrobes, and who are we to blame them?

The good news is that there is an enormous hidden value-creation opportunity to tap. Water consumption in global apparel production could in many cases be lowered by as much as 50 %, energy consumption by one third or more, and the use of chemicals reduced by up to one fifth [26]. To assess how powerful this is, consider the case of chemicals, explained below (Spotlight 6). As a by-product of the corresponding modernization, CO_2 emissions, costs, and production time can all be lowered as well. Not all these costs are internalized of course. Water, for example, is in many cases free and not even metered, but energy and chemicals do cost money (and create externalities in the environment). Water that has to be heated or cooled does cost money as well. As the scarcity of fresh water moves up the development agenda, one can envision solutions to incentivize a more efficient use of water in the textile production process, where it can take up to 2700 l just to produce the cotton needed to make a single T-shirt [27]. Better chemicals, machines, process technology and materials innovation can jointly unlock massive value added and provide the economic basis for improving working conditions and environmental footprints. This if accompanied by more modern management techniques that engage with workers on the shop floor. To seize this potential and take manufacturing to the next level, we need to start treating the sustainability dimension as the key driver of success it could be.

To be sure, referring back to the overwhelming majority of corporate CEOs in Accenture's 2010 research study on corporate sustainability—the 93 % who responded that sustainability would be critical to the future success of their companies; and this despite the then-recent economic downturn—one might ask why we are not there yet [28]. A 2013 follow-up to the Accenture survey revealed that many business leaders "have found themselves stuck on their ascent" to sustainability, "unable to scale sustainability at the pace required to address global challenges." Respondents to the 2013 edition of the survey "described a plateau beyond which they cannot progress without radical changes in market structures and systems, driven by a common understanding of global priorities" [29]. Part I examined how to engineer future-oriented business innovation in large corporate supertankers. Right now the orchestra performing on the Titanic would like to move beyond arranging the deck chairs—but is unsure how.

The following chapters ask how stakeholders in the apparel industry can achieve the relevant leap forward. They leverage insights in innovation and market transformation gathered over a number of years. Further, they are informed by evidence gathered from a yearlong study on the textile and garment industry I led in 2013–2014. We reached out to over 730 stakeholders in the industry through an online survey and conducted numerous expert interviews and site visits over the course of 2013, as well as reviewing over 200 reports on the overall industry or some relevant aspect of it.

The premise is simple: the key breakthrough ahead of us will be in determining how to finance greater sustainability on a long-term basis by achieving greater value added. This means assessing how fundamental shifts that are moving sustainability from the periphery to the core of value creation can be leveraged in the textile and garment industry. We need to determine how global supply chains can be harnessed to drive inclusive value creation in the context of trends such as green growth and total resource productivity and energy efficiency, the rise of the LOHAS consumer, as well as growing demand for affordable products and services at the Base of the Pyramid [30].

Raising the game will be hard work. It will require significant stakeholder engagement across the board. The silver lining on the horizon is that the road ahead for companies, civil society, and government need not be divergent. With the rise of Asia and fast fashion, the industry is already in the midst of transformation. A truly future proof industry that unlocks economic, environmental, and social value now rests with our ambition to take a longer view—and our determination to act ever bolder and design the systemic solutions.

Spotlight 6: Chemicals—Decisive for Worker Health and Safety

The apparel supply chain is very resource intensive. This also applies to the use of chemical inputs. According to Greenpeace, approximately 25 % of globally manufactured chemicals are used in the textile and garment industry [31]. Between 100 and 1000 g of chemicals are used to produce 1 kg of fabric [32]. Many of these chemicals have adverse effects on human beings. Exposure to chemical substances in textile products may result in considerable negative health outcomes, which range from acute poisoning to long-term consequences such as cancer. Accordingly, any strategy for engineering sustainable apparel supply chains must take into consideration the impact of the use of chemicals and their contribution to worker health and safety objectives, consumers, and the environment in general [33].

Moreover, though tightening, global regulation is uneven. With the introduction of stricter regulation in Europe via codified Registration, Evaluation and Authorisation of Chemicals (REACH) in 2007, many chemical producers have moved their production sites to Asia [34]. China has become the largest market for textile chemicals, as evident in ever-growing apparel and textile production and huge production of synthetic fibers and cotton [35]. China's textile industry now uses about 42 % of the world's textile chemicals [36]. In fact, most of the industry growth in the past 25 years has been driven by Asia, which now receives almost half of global chemical sales [37]. REACH-like legislation has also come under discussion in Asia, namely in Japan with the amendment in May 2009 of the "Chemical Substances Control Law" of 1973. In China the proposed amendment in May 2009 of the "Measures for the Environmental Administration of New Chemical Substances" of 2003 raised the bar, and paved the way for the implementation of a similar legislation as in the EU, also referred to as China REACH. Comparable processes also took place in Taiwan with the draft "Guidelines for Existing Chemical Substance Nomination and New Chemical Notification" announced in April 2009, and in India where a consultation process of the draft "National Chemicals Policy" began in 2012 [38].

Several things can be done regarding chemicals in the production process. First, process optimization can reduce the amount of chemicals required to manufacture a given product, often generating savings of up to 20 % [26]. Substituting hazardous chemicals for other agents in the production process that are less dangerous is also a viable tactic. Some chemicals used for textile production are known to have intrinsic hazardous properties that make their release into the environment of particular concern. A large number have never been properly tested for their safety [39].

Training should similarly be taken seriously. After the offshoring of production to emerging markets, textile workers are regularly insufficiently trained in how to work with hazardous chemicals. Further, they typically lack adequate protection when handling dangerous substances. The value creation logic of the industry after the advent of fast fashion is important to understand when seeking to upgrade environmental performance and worker health and safety. The short timeframe inherent to producing fast fashion (i.e., only about 14 days for a fast fashion retailer such as Inditex/Zara) means that there is no time to systematically test the end-product. Researchers found that the residue of a variety of hazardous chemicals was present in clothing made by 20 global fashion brands [40].

In addition, not all chemicals used during the manufacturing process leave traces on the endproduct shipped. This despite their having a substantial impact on the environment through water discharge at the production site. Selecting the least hazardous chemical inputs rather than focusing on testing the output product is one potential way forward. As chemical suppliers do not publicly disclose the composition of all chemicals used in the production process for reasons of confidentiality, industry upgrading will be difficult to achieve without the involvement of these players [33].

On top of efforts directly targeting worker and consumer health and safety, a logical early step is to systematically quantify the cost savings potential resulting from optimized processes. This shows the economic upside of better practice, and can help disseminate information about the shelf of comparatively safe chemical inputs. One necessary condition to achieve progress is the adoption of a front-of-pipe perspective, which is the focus of Swiss firm Bluesign Technologies. Founded in 2000 by a group of textile and chemical experts and controlled by Société Générale de Surveillance (SGS), a global certification company, Bluesign Technologies operates a label that begins at the front of the pipe. For its system partners, it assesses chemical inputs that have an enormous impact on environmental pollution and worker health and safety. Bluesign utilizes a comprehensive process—in partnership with the chemical industry, brands, textile and garment suppliers—that creates transparency mainly about environmental impacts, and helps define appropriate processes and provide a screened shelf of chemical inputs. In 2015, Bluesign unveiled its blueXpert tool. Developed with leading chemical businesses Archroma, Dystar, and Huntsman, the online tool seeks to enable textile mills to benchmark the efficiency of their wet processing operations against optimized best industry practices, thereby enabling them to significantly improve their total resource productivity [41].

Such a chemical change management process can be the catalyst to significantly increase total resource productivity and can lead to significant cost reductions down the road. A front-of-pipe approach that operates with optimized production processes and a pre-approved shelf of chemicals that minimize environmental impacts is the only way to have environmentally sustainable processes in place when orders scheduled for fast turnaround hit the producer.

4.2 Global Fashion: New Buyers, New Producers

The key development of the past 30 years is the emergence of buyer-led global supply chains. They produce textiles and garments in ever accelerating design, production and distribution cycles. To grasp upgrading opportunities in the apparel industry, let us first consider the key features within its value chain, as visualized in the figure below. The value chain ranges from the production of raw materials (i.e., natural as well as manufactured fibers) to the manufacture of a wide variety of semi-finished and finished products [42].

Downstream parts of the textile and garment industry, such as the clothing industry, consume the output of the upstream parts such as fabrics of all types and colors. Inputs from the agricultural sector, such as natural fibers like cotton or wool, are used in the production of textiles and garments. The chemical industry provides thousands of chemicals for the textile and garment industry, such as artificial fibers and chemicals used for bleaching, dyeing, mercerizing, printing or softening [42]. Below is a schematic view of the textile and garment value chain (Fig. 4.1).

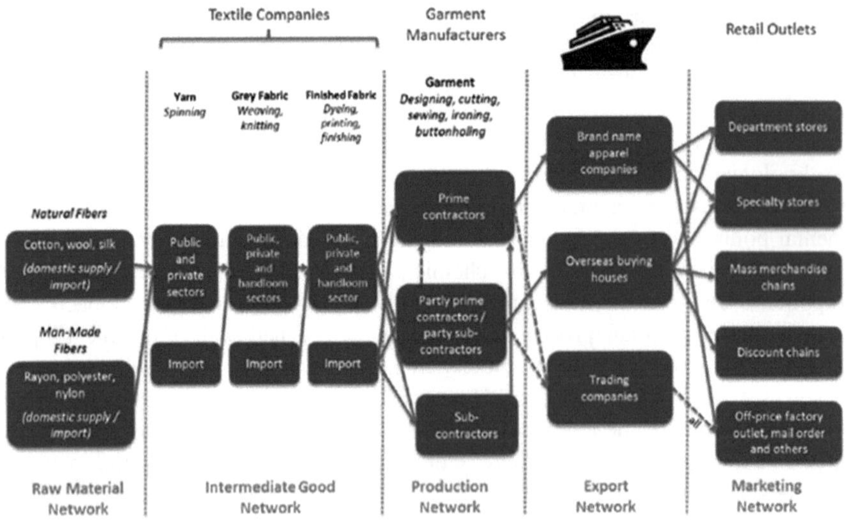

Fig. 4.1 The apparel value chain (*source* author) [43, p. 6]

One of the reasons that resource-optimizing production is not standard in the industry relates to offshoring to locations with, in many cases, lower availability of skilled labor and significantly less stringent local environmental and labor requirements. The many shifts in the locations of the most significant apparel exporting countries and regions and their main end markets have been driven by preferential tariff treatment and cost; other factors are proximity to consumer markets and ability to provide unique value adds [8].

In 2014, the world's largest apparel consumer markets were the European Union, the US, and Japan. Apparel import penetration was generally very high, but varied significantly across the main consumer countries. In 2012, the European Union (EU-27, including intra-EU-27 trade) accounted for approximately 40.2 % of total world apparel imports of USD 423 billion, while the US accounted for 20.8 %, and Japan for 8.0 %. The US, the EU-27, and Japan together represented over two-thirds (69 %) of world apparel imports in 2012, down from 82.4 % in 1995. Notably, the US's share of global apparel imports steadily declined from a peak of 33.9 % in 2000 to 20.8 % in 2012. Japan dropped similarly from 11.5 % in 1995 to 8 % in 2008 [44]. With a retail market worth more than USD 120 billion Japan was nevertheless one of the biggest players in the global textile and garment industry [45]. Moreover, Japan has traditionally set the bar in terms of high quality textile and garment production facilities—the quality of certain products needed in high-end Western brands is so high that they source from Japan, regardless of price. Despite that, there was little investment into the local industry in the 2010s, which was going through the process of being offshored.

In 2013, the world's largest textile and clothing producer countries were China, EU-28 (Croatia had joined the EU in the meantime), India, Turkey and Bangladesh [46]. Vietnam was sixth and growing fast. With the Trans-Pacific Partnership (TPP) around the corner—a free-trade agreement that could halve or even nullify the 17 % duty the US was levying on garment imports from Vietnam—, the country's garment industry is likely to grow even faster soon (this raising some questions, as e.g. a mere 42 of 639 industrial clusters had wastewater treatment systems in 2015) [47].

For now, a period of tremendous growth has turned Bangladesh into Asia's number two sourcing hotspot, though as a distant second to China. By 2014, the industry was the country's main driver of GDP growth and accounted for about 20 % of GDP in Bangladesh, 80 % of total export earnings, and over 4 million direct jobs [48–50]. Looking ahead to the golden jubilee of independence, Bangladesh aimed to raise its exports from USD 24.5 billion in 2013/14 to USD 50 billion by 2021 [51, 52]. More generally, South Asia's global exports of textiles and clothing increased substantially over the past decade. South Asia has been one of the leading garment manufacturing locations in terms of in-country contribution to output, employment and trade. South Asian countries, specifically Bangladesh, India, Indonesia, Vietnam, Pakistan and Sri Lanka, compete fiercely with one another in the global market. They now account for three-quarters of world clothing exports [44, 51, 52].

Spotlight 7: China—King of Garments, Emperor of Luxury?
The apparel market in China is evolving, and is on its way to becoming the biggest apparel market in the world. With a 35 % share in global textile exports, and a 39 % share in world clothing exports, China also remained the world's leading exporter in 2013, exporting USD 106.6 billion worth of textiles, and USD 177 billion worth of clothing [53, 54]. The changes occurring in the apparel industry in China are tied in large part to the broader social and economic context of the country. The Chinese have experienced tremendous economic growth and urbanization over the past sixty years, with rapidly growing middle- and upper-classes that are projected to encompass 75 % of the Chinese population by 2025 as a result [55]. These consumers—who demand different types of products, including those associated with the LOHAS segment—are projected to continue to rise [55]. Additionally, demand at the Base of the Pyramid and beyond just the wealthiest Chinese consumers is swelling at increasingly fast rates, as hundreds of millions of Chinese step out of poverty. The textile and garment industry, a highly fragmented industry in China with many different producers beyond just state-owned companies, is ripe with opportunity, particularly given the need for consolidation. As a leading executive remarked, "The lack of big companies in the industry is an opportunity" [56]. Surveying Chinese consumers revealed that they were nearly twice as familiar in 2010 with the names of ready-to-wear brands as they were even in 2008 [57]. A 2012 survey by KPMG confirmed that brand recognition continues to grow among consumers, as does the importance these consumers place on the luxury variety [58]. Chinese consumers have also become "savvier about the relationship between quality and price" and that internationally well-known brands have "become one of the most important factors in making a purchase" [57].

What this means for foreign firms looking to penetrate the Chinese market is twofold. First, China's luxury-goods market is growing. It is estimated to account for more than EUR 74 billion of global luxury sales by 2020 (44 %) [59]. This is not a market that can risk being ignored. Nor is it one where first movers should chance being unresponsive to the tastes of Chinese consumers, who are particularly comfortable with displays of wealth and the purchasing of luxury goods. The pressure for foreign companies to act sooner rather than later is growing, principally because the Chinese government announced a "Going Out" strategy as part of the country's Twelfth Five Year Plan [60]. The plan focuses on penetrating foreign markets as a way to expand markets for selling and producing goods in light of domestic issues such as excess capacity, rising costs, domestic demand limitations, environmental restrictions, appreciation of the Chinese currency, and inadequate infrastructure [60].

Second, China has raised the bar especially on the industry's environmental impact, and growth is expected to be flat. In 2010, the Ministry of Environmental Protection (MEP) introduced a new chemicals regulation similar to EU REACH, called China REACH [61]. In 2013, China acknowledged the existence of so-called 'cancer villages' for the first time, resulting from textile production [62]. The 2011–2015 Five Year Plan referred to the risks that toxic chemicals pose to the country's environment and the health of its population, listing seven industries including the textile industry for "priority pollution prevention control." The plan blacklisted 58 chemicals and targeted an elimination chemical list by 2015 [62].

Mainly responding to the Greenpeace Detox campaign, a number of major apparel and footwear brands and retailers committed to work towards zero discharge of hazardous chemicals (ZDHC) by 2020. In China, ZDHC initiated a collaboration with the China National Textile and Apparel Council and conducted chemicals management training for suppliers [63]. In 2015, the Chinese government zoomed in further, now on water: it handed out more fines to put pressure on garment manufacturers to reduce water pollution in China, while global brands asked their suppliers to improve wastewater treatment and recycling [64].

4.3 Social and Environmental Issues We Need to Solve

Given its ability to absorb unskilled labor into formal employment, the general importance of the textile and garment industry for development cannot be overestimated. For countries that have experienced an industrial revolution, the apparel industry has almost always spearheaded the shift [9]. It also provides a context where relatively modern technology can be deployed at low investment cost, driving development dividends through backward and forward linkages with the economy [65]. Confirming this, 82.6 % of respondents to the 2013 Impact Economy online survey ranked "providing an opportunity to work and gain a salary" among the top three positive contributions of the global textile and garment industry for workers. Notwithstanding, looms and other machinery are often 40 years or older. Deploying newer equipment can create tremendous productivity and efficiency gains.

Often a source of economic progress, in developing countries apparel also has a track record of poor working conditions and heavy pollution. The social environment in the sourcing countries is often inherently unstable and sometimes downright volatile. Political unrest in Bangladesh cost the country USD 2.2 billion in lost GDP in the first quarter of 2015 [66]. Labor rights, wage levels and worker health and safety are the leading social issues to be sorted out. Understood in the broader sense to include child labor, forced labor and migrant labor, labor rights are probably the predominant issue in the sourcing countries [67].

Wage levels are another high priority and contentious topic. There is intensifying discussion about living wages. The Bangladeshi parliament passed the Bangladesh Labour (Amendment) Bill on July 15, 2013 in response to domestic and international pressure. Before, Bangladesh paid the world's lowest minimum wages for garment workers, at about USD 38 a month. The minimum monthly wage for the country's garment workers was raised by 77 % to Bangladeshi Taka (BDT) 5300 (USD 68) effective December 1, 2013, following numerous workers' protests that forced about 250 garment factories to close temporarily [68]. To illustrate how much new thinking is in the air, consider Cambodia: in September 2014, eight fashion brands—including C&A, H&M, Inditex, Primark, Next, New Look, and Tchibo—wrote to the prime minister and the Garment Manufacturers Association in Cambodia (GMAC) arguing for a *higher* minimum wage for Cambodia's 600,000 garment workers, aiming for "a fair living wage" [69]. New tools to calculate a living wage have also come on stream. In 2015, the Ethical Trading Initiatives (ETI) of Denmark, Norway and the UK published a report aiming at repositioning the debate on living wages for companies within global supply chains [70]. The recommendations went beyond just looking at a company's context, asking how to set an acceptable level in the brand's own supply chain, and then get suppliers to pay. Instead, the three ethical trading initiatives argued that brands should coordinate between themselves, and collaborate "with suppliers, employers' associations, trade unions, NGOs and national governments, including in relation to setting adequate national minimum wages" [70]. An even bigger departure from standard practice, the ETIs call on companies to actively support collective bargaining, including the emergence of functioning local collective bargaining mechanisms and institutions, especially trade unions. Notwithstanding, the only way to raise wages systematically without simply setting in motion a wave of relocation and subcontracting is to raise productivity. The logical goal in many producer countries is to break out of the low skill/low wage strategy.

Moreover, occupational health and safety (OHS) remains a key focal point in the industry, dramatically showcased by the Rana Plaza accident, as well as other accidents that have happened since, albeit on a lesser scale. According to the ILO, work-related accidents and disease continue to be serious problems. They are caused by a wide range of endemic factors, including a widespread lack of compliance with health and safety standards at factories [9, 71]. A lot of work lies ahead on this front. In Bangladesh, a vast majority of the more than 5600 factories producing for export need upgrading. It is good news that the majority of them is being assessed in terms of structural, electrical and fire safety. However, the sector is growing. Simply extrapolating on the trend, there could be more than 10 million workers and over 10,000 factories in the country by the golden jubilee in 2021, unless significant factory consolidation takes place. The case of Bangladesh is discussed below (Spotlight 8).

Social and environmental issues are the bleeding sustainability edge of the textile and garment industry. But looking at the organizational capacity of the leading actors working on labor rights, living wages and health and safety, it is difficult to imagine how their already insufficient capacity to address these issues could grow in parallel without any structured effort to render some of them larger and more effective. Additional capacity needs to be brought on stream to be able to improve social and environmental performance. A mindset shift towards understanding that sustainability can be a source of economic benefits is a necessary condition to generating this extra bandwidth. One expert responding to the Impact Economy survey put it well, "Constant improvement of performance regarding surface problems like worker safety and worker rights should be seen as a competitive advantage" [65].

Spotlight 8: Bangladesh—From Fast-Only to Sustainable Growth?
In Bangladesh, the textile and garment industry has become a major contributor to development since its beginnings in the mid-1980s. Apparel has been instrumental in reducing poverty by a third since the 1990s. Accidents such as Rana Plaza serve as a constant reminder that globalization has a dark side as well. Bangladesh has also emerged as the largest production cluster in South Asia, with the exception of China. With over 5600 RMG factories and an average of 1500–2000 workers per factory, Bangladesh was far ahead of its main regional competitors, namely Indonesia with about 2450 RMG factories, Vietnam with about 2000, and Cambodia with 260 in 2012, though Vietnam in particular has been growing fast and seeks to close the gap [72–74]. The Bangladeshi market cluster extends across a geographically small area. This facilitates transportation and subcontracting. This advantages Bangladesh vis-à-vis other volume producers such as India and Pakistan [72–74]. The growth record and projections for the industry in Bangladesh are impressive. Yet, views on Bangladesh's potential vary widely. Some, such as former Goldman Sachs chief economist Jim O'Neill, believe the densely populated and youthful country has the potential to become one of the world's most vibrant economies. He includes Bangladesh in his "Next Eleven" (N-11) group of countries that he expects to lead the next wave of high-growth economies that could potentially have a Brazil, Russia, India, China (BRIC)-like importance. J.P. Morgan sees Bangladesh as part of the "Frontier Five" key frontier markets [75, 76]. Others see Bangladesh's development potential blocked by political infighting and overstretched infrastructure, with business hindered by red tape, corruption, energy shortages and the ever-present risk of social unrest. Bangladesh could possibly even become a "climate prison" once rising sea levels will eventually reduce the size of the country by up to 17 % [77].

For Bangladesh to become a high-growth economy that benefits the population at large, the gap between reality and ambition needs to be bridged. Well-placed measures are needed that take the industry's social and environmental performance to the next level in ways that do not undercut its competitiveness. As one respondent to the Impact Economy survey remarked, "If international help is provided to Bangladesh in the form of Generalized System of Preferences (GSP), the country can retain the same growth rate, and the export rate would cross USD 30 billion within the next five years" [65]. What's more, apparel production in Bangladesh is positioned quite competitively in the region. The country has the capacity to increase the quality of goods produced (particularly as Japan and Thailand increasingly offshore production), while still remaining competitive with emerging production clusters such as Myanmar. Vietnam will increasingly serve the US market. More growth and development will however test the limits of already severe infrastructure deficits. Sustainable supply chains with corresponding upgrading of physical infrastructure, human capital and higher productivity will be key to overcoming these limitations, so that Bangladesh can grow fast and sustainably.

References

1. Ditty S (2015) Europe in the world: the garment, textiles and fashion industry. In: Fashion revolution. https://europa.eu/eyd2015/en/fashion-revolution/posts/europe-world-garment-textiles-and-fashion-industry. Accessed 14 Jul 2015
2. Greenpeace (2014) A red card for sportswear brands: hazardous chemicals found in World Cup merchandise. http://www.greenpeace.org/international/Global/international/publications/toxics/2014/Detox-Football-Report.pdf. Accessed 14 Jul 2015
3. European Chemicals Agency (2013) Draft background document for N,N-Dimethyl-formamide. http://echa.europa.eu/documents/10162/13640/5th_recom_draft_backgdoc_dmf_en.pdf. Accessed 14 Jul 2015
4. Smith N (2015) Taking PU waterborne: how to address the sustainability challenges of a key material in our industry. PowerPoint presented at the 4th Bluesign conference, St. Gallen, Switzerland, 13–14 Jul 2015. http://www.bluesign.com/sites/fourth-bluesign-conference/downloads#.VbeOeLqyM3s. Accessed 28 Jul 2015
5. Ghaly EA et al (2014) Production, characterization and treatment of textile effluents: a critical review. J Chem Eng Process Technol. http://www.omicsonline.org/open-access/production-characterization-and-treatment-of-textile-effluents-a-critical-review-2157-7048.1000182.pdf. Accessed 14 Jul 2015
6. MarketLine (2012) Global textiles, apparel and luxury goods. http://store.marketline.com/Product/global_textiles_apparel_luxury_goods?productid=MLIP0468-0001. Accessed 14 Nov 2013
7. Fernandez-Stark K et al (2011) The apparel global value chain. Duke Center on Globalization, Governance & Competitiveness. http://www.cggc.duke.edu/pdfs/2011-11-11_CGGC_Apparel-Global-Value-Chain.pdf. Accessed 14 Nov 2013
8. Gereffi G, Memedovic O (2003) The global apparel value chain. United Nations Industrial Development Organization. http://www.unido.org/fileadmin/media/documents/pdf/Services_Modules/Apparel_Value_Chain.pdf. Accessed 14 Nov 2013
9. Islam M et al (2013) Textile industries in Bangladesh and challenges of growth. Res J Eng Sci. http://www.isca.in/IJES/Archive/v2i2/7.ISCA-RJEngS-2013-002.pdf. Accessed 25 Oct 2013
10. British Broadcasting Corporation–BBC (2013) Dhaka factory collapse: no compensation without DNA identification. http://www.bbc.co.uk/news/magazine-24080579. Accessed 14 Nov 2013
11. Labor Rights (2013) The Labor Rights register. http://www.laborrights.org/creating-a-sweatfree-world/resources/bangladesh-fire-and-building-safety-agreement. Accessed 14 Nov 2013
12. Alliance for Bangladesh Worker Safety (2013) The Alliance for Bangladesh Worker Safety register. http://www.bangladeshworkersafety.org. Accessed 14 Nov 2013
13. Morris S (2013) Bangladesh factory collapse: clothes made for a tenth of retail price, documents show. The Star. http://www.thestar.com/news/world/2013/05/14/bangladesh_factory_collapse_clothes_made_for_a_tenth_of_retail_price_documents_show.html. Accessed 15 Jul 2015
14. Leonie BL (2014) Viewpoint: cotton price decline unlikely to help clothing costs. http://www.just-style.com/comment/cotton-price-decline-unlikely-to-help-clothing-costs_id123070.aspx. Accessed 15 Jul 2015
15. Clifford S (2013) That 'Made in U.S.A.' premium. New York Times. http://www.nytimes.com/2013/12/01/business/that-made-in-usa-premium.html
16. Falguni S (2014) An overview of NIKE's supply chain and manufacturing strategies. http://marketrealist.com/2014/12/overview-nikes-supply-chain-manufacturing-strategies/. Accessed 15 Jul 2015
17. Bhardwaj V, Fairhurst A (2010) Fast fashion: response to changes in the fashion industry. http://www.tandfonline.com/doi/pdf/10.1080/09593960903498300. Accessed 14 Nov 2013

18. Retail-Assembly.org (2013) H&M's history of designer collaborations. http://retail-assembly.org/blog/hms-history-of-designer-collaborations. Accessed 14 Nov 2013
19. CLSA (2011) Dipped in gold: luxury lifestyles in China. https://www.clsa.com/about-clsa/media-centre/2011-media-releases/china-to-become-the-worlds-largest-market-for-luxury-goods.php. Accessed 14 Nov 2013
20. Bangladesh Accord on Fire and Building Safety released (2013). The Bangladesh Accord on Fire and Building Safety released register. http://www.industriall-union.org/bangladesh-accord-on-fire-and-building-safety-released. Accessed 14 Nov 2013
21. Alliance for Bangladesh Worker Safety (2013) The Bangladesh Worker Safety register. http://www.bangladeshworkersafety.org. Accessed 14 Nov 2013
22. Accord on Fire and Building Safety in Bangladesh (2015) The Bangladesh Accord register. http://bangladeshaccord.org/progress/. Accessed 14 Jul 2015
23. Historical Boy's Clothing (2013) World War II rationing. http://histclo.com/mat/rat/rat-ww2.html. Accessed 14 Nov 2013
24. Koch W (2013) Clothes recycling goes curbside as demand rises. USA Today. http://www.usatoday.com/story/news/nation/2013/04/20/recyling-clothes-expands-curbside/2092351/. Accessed 14 Nov 2013
25. University of Cambridge Institute for Manufacturing (2006) Well dressed? The present and future sustainability of clothing and textiles in the United Kingdom. http://www.ifm.eng.cam.ac.uk/uploads/Resources/Other_Reports/UK_textiles.pdf. Accessed 14 Nov 2013
26. Impact Economy (2013) Expert interview for sustainable apparel value chains project, Geneva
27. World Wildlife Fund (2013) The impact of a cotton T-shirt: how smart choices can make a difference in our water and energy footprint. http://www.worldwildlife.org/stories/the-impact-of-a-cotton-t-shirt. Accessed 14 Jul 2015
28. UN Global Compact-Accenture (2010) A new era of sustainability: UN Global Compact-Accenture CEO study 2010. http://www.unglobalcompact.org/docs/news_events/8.1/UNGC_Accenture_CEO_Study_2010.pdf. Accessed 14 Nov 2013
29. UN Global Compact-Accenture (2013) The UN Global Compact-Accenture CEO study on sustainability 2013. http://www.accenture.com/Microsites/ungc-ceo-study/Documents/pdf/13-1739_UNGC%20report_Final_FSC3.pdf. Accessed 14 Nov 2013
30. Martin M (2013) CSR's new deal: a blueprint for your first hundred days in the world of impact economy. Impact Economy. http://www.impacteconomy.com/CSR_new_deal_russian. Accessed 14 Nov 2013
31. Greenpeace (2013) Chemicals substitution in textile industry: implementing chemical policies into the textile supply chain. http://www.greenpeace.org/china/Global/china/publications/others/2013/20130325-GP-ENG.pdf. Accessed 14, 29 Nov 2013
32. Lacasse K, Baumann W (2004) Textile chemicals: environmental data and facts. German Environmental Protection Agency, p 609. http://oecotextiles.wordpress.com/2013/01/10/chemicals-used-in-textile-processing. Accessed 29 Nov 2013
33. Swedish Chemicals Agency (2013) Hazardous chemicals in textiles. http://www.kemi.se/Documents/Publikationer/Trycksaker/Rapporter/Rapport-3-13-textiles.pdf. Accessed 2 Dec 2013
34. The new European chemicals regulation REACH (2007) came into force on June 1, 2007 aiming at improving protection of human health and the environment. European Commission (2015) The European Commission register. http://ec.europa.eu/environment/chemicals/reach/reach_en.htm. Accessed 15 Jul 2015
35. IHS Chemical (2010) Textile chemicals. http://www.ihs.com/products/chemical/planning/scup/textile.aspx. Accessed 2 Dec 2013
36. Greenpeace (2013) Chemicals substitution in textile industry: implementing chemical policies into the textile supply chain. http://www.greenpeace.org/china/Global/china/publications/others/2013/20130325-GP-ENG.pdf. Accessed 29 Nov 2013

37. AT Kearney (2012) Chemical industry vision 2013: a European perspective. http://www.atkearney.com/documents/10192/536196/Chemical+Industry+Vision+2030+A+European+Perspective.pdf/7178b150-22d9-4b50-9125-1f1b3a9361ef. Accessed 2 Dec 2013
38. ICIS (2010) REACH-like regulations enacted globally: a regulatory world tour. http://www.icis.com/Articles/2010/05/31/9362538/reach-like-regulations-enacted-globally.html. Accessed 5 Dec 2013
39. Greenpeace (2009) Hazardous chemical pollution of the Pearl River: technical note 08/2009. http://www.greenpeace.org/international/Global/international/planet-2/report/2009/10/hazardous-chemical-pollution-o.pdf. Accessed 2 Dec 2013
40. Greenpeace (2012) Toxic threads: the big fashion stitch-up. http://www.greenpeace.org/international/Global/international/publications/toxics/Water%202012/ToxicThreads01.pdf. Accessed 2 Dec 2013
41. Brett M (2015) Bluesign launches blueXpert online tool. Ecotextile. http://www.ecotextile.com/2015071321591/dyes-chemicals-news/bluesign-launches-bluxpert-online-tool.html. Accessed 16 Jul 2015
42. Stengg W (2001) The textile and clothing industry in the EU. http://www.viva-systems.com/english/enterprise_paper_02_2001.pdf. Accessed 14 Nov 2013
43. Martin M (2013) Creating sustainable apparel value chains. A primer on industry transformation, p 6. Impact Economy. http://www.impacteconomy.com/papers/IE_PRIMER_DECEMBER2013_EN.pdf. Accessed 15 Jul 2015
44. WTO (2013) The WTO register. http://stat.wto.org/Home/WSDBHome.aspx?Language. Accessed 14 Nov 2013 (Apparel represented by SITC Code 84)
45. Hirono T (2011) Japan: contemporary women's wear market. US Commercial Service, US Department of Commerce. https://www.nccommerce.com/Portals/5/Documents/ITD/Japan%20Contemporary%20Women%E2%80%99s%20Wear%20Market.pdf. Accessed 15 Jul 2015
46. WTO (2014) Europe's exports of manufactured goods totalled almost US$ 5 trillion in 2013. http://www.wto.org/english/res_e/statis_e/its2014_e/its14_highlights2_e.pdf. Accessed 15 Jul 2015
47. Ecotextile News (2015) Perfect storm. Ecotextile News 67(Jun/Jul): 40–41
48. Asian Development Bank (2013) Quarterly economic update on Bangladesh. http://www.adb.org/sites/default/files/pub/2013/ban-qeu-2013-03.pdf. Accessed 14 Nov 2013
49. Bangladesh Garment Manufacturers and Exporters Association-BGMEA (2013) Trade information 2012/2013. http://bgmea.com.bd/home/pages/TradeInformation#.UgH9sZL0FRs. Accessed 25 Oct 2013
50. BOI Bangladesh, Prime Minister's Office (2013). Key statistics of RMG sector. http://www.boi.gov.bd/index.php/potential-sector/garments-and-textitles. Accessed 25 Oct 2013
51. FashionUnited (2014) Bangladesh sets export target of 50 billion US dollars by 2021. https://fashionunited.uk/news/business/bangladesh-sets-export-target-of-50-billion-us-dollars-by-2021/2014121514895. Accessed 15 Jul 2015
52. Reuters (2014) Bangladesh exports up 11 pct in 2013/2014 as garment sales surge. http://www.reuters.com/article/2014/07/10/bangladesh-economy-exports-idUSL4N0PL25P20140710. Accessed 15 Jul 2015
53. Tara D (2014) WTO: textile and clothing exports surpass average growth. Sourcing Journal Online. https://www.sourcingjournalonline.com/wto-textile-clothing-exports-surpass-average-growth-td. Accessed 15 Jul 2015
54. WTO (2013) Apparel SITC Code 84, Textiles SITC Code 65. http://stat.wto.org/Home/WSDBHome.aspx?Language; Eurostat. Accessed 14 Nov 2013
55. World Business Council for Sustainable Development (2013) Vision 2050: the new agenda for business. http://www.wbcsd.org/pages/edocument/edocumentdetails.aspx?id=219. Accessed 14 Nov 2013
56. Flannery R (2013) Where's China's growth? Textile industry is weaving expansion. Forbes. http://www.forbes.com/sites/russellflannery/2013/08/28/wheres-chinas-growth-check-out-textiles-texhong-and-hong-tianzhu/. Accessed 12 Nov 2013

57. Atsmon Y et al (2010) Tapping China's luxury-goods market. McKinsey & Company, New York. http://www.mckinsey.com/insights/marketing_sales/tapping_chinas_luxury-goods_market. Accessed 12 Nov 2013
58. KPMG (2013) Global reach of China luxury. https://www.kpmg.com/CN/en/IssuesAndInsights/ArticlesPublications/Documents/Global-Reach-China-Luxury-201301.pdf. Accessed 12 Nov 2013
59. CLSA (2011) China to become the world's largest market for luxury goods over the next decade. https://www.clsa.com/about-clsa/media-centre/2011-media-releases/china-to-become-the-worlds-largest-market-for-luxury-goods.php. Accessed 18 May 2015
60. Fibre2Fashion (2013) Chinese textile sector discusses 'Going Out' strategy. http://www.fibre2fashion.com/news/textile-news/newsdetails.aspx?news_id=148034. Accessed 12 Nov 2013
61. Chemical Inspection & Regulation Service (2014) New chemical substance notification in China—China REACH. http://www.cirs-reach.com/China_Chemical_Regulation/IECSC_China_REACH_China_New_Chemical_Registration.html. Accessed 15 Jul 2015
62. Fashion United (2013) China, the garment king: a portrait. https://www.fashionunited.com/executive/management/china-the-garment-king-a-portrait-20132310493259. Accessed 15 Jul 2015
63. Zero Discharge of Hazardous Chemicals Programme-ZDHC (2015) 2014 annual report and 2013 annual report. http://www.roadmaptozero.com/programme-documents.php. Accessed 15 Jul 2015
64. Chu K (2015) Chinese apparel makers face pressure to reduce water pollution. Wall Street J. http://www.wsj.com/articles/chinese-apparel-makers-face-pressure-to-reduce-water-pollution-1433301390. Accessed 2 Jun 2015
65. Impact Economy (2013) Online survey for sustainable apparel value chains project, Geneva
66. Bdnews24.com (2015) Political turmoil cost Bangladesh $2.2 billion in three months of 2015: World Bank. http://bdnews24.com/economy/2015/04/12/political-turmoil-cost-bangladesh-2.2-billion-in-three-months-of-2015-world-bank. Accessed 15 Jul 2015
67. The ILO Declaration on Fundamental Principles and Rights at Work lists four core labor standards for universal application, which are: (1) freedom of association and "effective recognition" of the right to collective bargaining; (2) elimination of forced labor; (3) effective abolition of child labor, and (4) elimination of discrimination in employment. International Labour Organization—ILO. The ILO register. www.ilo.org. Accessed 15 Jul 2015
68. Devnath A (2013) Bangladesh raises minimum wage for garment workers after unrest. http://www.bloomberg.com/news/2013-11-13/bangladesh-garment-factories-to-stay-shut-amid-worker-protests.html. Accessed 15 Nov 2013
69. Robertson H et al (2014) Global brands pledge to pay more for garments. Cambodia Daily. https://www.cambodiadaily.com/news/global-brands-pledge-to-pay-more-for-garments-68261. Accessed 15 Jul 2015
70. Dansk Initiativ for Etisk Handel, Ethical Trading Initiative Norway and Ethical Trading Initiative UK (2015). The ETI register. http://www.ethicaltrade.org/sites/default/files/resources/living-wages-in-global-supply-chains.pdf. Accessed 15 Jul 2015
71. ILO (2012) Bangladesh Decent Work Country Programme 2012–2015. http://www.ilo.org/public/english/bureau/program/dwcp/download/bangladesh.pdf. Accessed 25 Oct 2013
72. BGMEA (2013) Trade information 2012/2013. http://bgmea.com.bd/home/pages/TradeInformation#.UgJ_mZL0FRs. Accessed 25 Oct 2013
73. Fair Wear Foundation (2013) The Fair Wear register. http://www.fairwear.org/ul/cms/fck. Accessed 25 Oct 2013
74. McKinsey & Company (2012) Bangladesh's ready-made garments landscape: the challenge of growth. http://www.mckinsey.de/sites/mck_files/files/2011_McKinsey_Bangladesh.pdf. Accessed 25 Oct 2013
75. Goldman Sachs (2007) BRIC's and beyond. http://www.goldmansachs.com/our-thinking/archive/archive-pdfs/brics-book/brics-full-book.pdf. Accessed 25 Oct 2013

76. Morgan JP (2007) Frontier five: Ho Chi Minh trail to Mexico. http://at-capital.com/at/JP%20Morgan%20Frontier%20Five.pdf. Accessed 25 Oct 2013
77. Spangler C et al (2014) Borrowed time on disappearing land. New York Times. http://www.nytimes.com/2014/03/29/world/asia/facing-rising-seas-bangladesh-confronts-the-consequences-of-climate-change.html?_r=0. Accessed 15 May 2015

Changing Consumption and Production Patterns

5

5.1 Two Tribes: Fast Fashion Versus Virtuous Consumers

Driven by the desire to be the first to hold a new piece of affordable fashion in their hands, over 3000 shoppers stormed the opening of a new Primark store on Oxford Street in London in April 2007—injured staff, trampled shoppers, and stunned security guards emerged in their wake [1]. In many industries, sustainability issues are transforming industries by moving mainstream. Not so in the textile and garment industry (for now). The success of several ethical clothing brands, such as Patagonia, is both a testament to the power of consumer demand and green credentials, as it is a reminder of what makes a niche different from the mainstream [2].

Sustainability considerations have begun to raise the stakes in the industry as well, mirroring a broader trend. Notwithstanding, there is little indication thus far that consumer demand in the textile and garment industry alone will affect the production process upstream. This is an important consideration to be aware of as we consider the prospect of industry transformation. For now it seems that sustainable consumer fashion is a small niche, despite many laudable attempts to educate the public. Most consumers want fast fashion at cutthroat prices rather than demanding better wages for workers and safer working conditions [3]. Buyers tend to source fashion-sensitive products from suppliers that can deliver in a flexible and speedy manner. Basic products are sourced from countries with the lowest costs. This dynamic is more complicated than a first glance would suggest, given the differences between fast fashion and quick response. Fast fashion emerged from quick response but they are distinct concepts. Quick response is associated with replenishment purchases for basic products. Fast fashion is quick response in new merchandise (with little or no replenishment), involving shipping fewer pieces in a great variety of styles, and at a higher frequency. In both cases, the volatility in sourcing volumes and required delivery speed can put significant stress on producers. They typically outsource part of the orders they get, often to "tier two" sweatshop style factories that are characterized by poor working conditions. Their role in the fulfillment of orders remains invisible to the buyer abroad.

M. Martin, *Building the Impact Economy*,
DOI 10.1007/978-3-319-25604-7_5

5.2 Transparency: Breathing Down the Supply Chain's Neck

While sustainable consumption practices will no doubt move mainstream over time, a much more compelling reason to now engage in industry transformation has to do with managing risk and securing efficiency gains. The roles and relationships among national and global lead firms, apparel manufacturers, and intermediaries in the textile and garment industry have become increasingly blurred in recent years. So-called "lead firms" include retailers and brand owners and are headquartered in the leading markets, namely Europe, Japan, and the United States. These firms tend to perform the most valuable activities in the apparel value chain, namely design, branding, and marketing of products. After the long recession in the main consumer markets resulting from the 2008 financial crisis, the intensity of competition has increased: "survival of the fittest" is the name of the game. Supply chain rationalization is a way to stay competitive and it is changing the face of the industry [3]. By "trimming the fat," lead firms are increasingly confining their relationships to their most capable and reliable suppliers and are significantly reducing the size and scope of their supply chains. Simplicity is king. Buyers want to deal with fewer, larger, and more capable suppliers, who are strategically located near major markets around the globe. As buyers seek to consolidate the number of wholesalers they source from, they demand a more comprehensive line of clothing, accessories, and footwear from these wholesalers [3].

Coinciding with supply chain rationalization and lean manufacturing, attention to sustainability and transparency in business is advancing along a broad front. Growing consumer demand for higher social and environmental standards across the board has increased the need for supply chain transparency in both the US and in Europe. This is affecting the textile and garment industry as well. Beyond emerging consumer consciousness, environmental compliance requirements and social initiatives in apparel are moving center stage. This is largely because of the work of advocacy groups and the role model effect of the more ambitious corporate social responsibility programs of some companies, such as Nike or Patagonia [4–6].

Important political efforts that change framework conditions are underway as well. At a local level, former California Governor Arnold Schwarzenegger signed "The California Transparency in Supply Chains Act of 2010" into law in September 2010 [7]. The legislation is intended to provide public information about the activities manufacturers engage in, and to monitor their supply chains to prevent human trafficking and modern forms of slavery. Disclosures allow businesses and consumers to make more informed decisions regarding the products they choose to purchase and the companies with whom they choose to conduct business. The role information technology (IT) can play in achieving transparency at the shop floor level is discussed below (Spotlight 9).

At the federal level in the US, the Obama administration expressed its concerns about abuses in the industry with the symbolic decision to suspend 37-year-old trade privileges that provided duty-free treatment with Bangladesh in June 2013, to pressure factories and brands to comply with safety standards [8, 9]. While the Generalized System of Preferences (GSP) affects less than 1 % of Bangladesh's exports of goods, as garment was never part of the preferential treatment, Obama's sanctions have added to the visibility of poor working conditions in the industry. In January 2015, the US Trade Representative recognized progress in its GSP Review of Bangladesh, but urged for more work on worker safety and rights prior to a reinstatement [10].

Transformation efforts have also been underway at the industry level for some time. For example, the Sustainable Apparel Coalition (SAC) was formed in 2011 as an industry-wide group of over 100 leading apparel and footwear brands, retailers, suppliers, nonprofits, and NGOs, working to reduce the environmental (and social) impacts of apparel and footwear products around the world [11]. The coalition puts useful tools at the disposal of its members to create transparency around product design and production decisions, especially in terms of their environmental impact.

Focusing on the social performance dimension, Fair Wear Foundation (FWF) is an international nonprofit verification initiative dedicated to enhancing workers' lives around the world [12]. Fair Wear Foundation is based in seven European countries and works with companies and factories to improve labor conditions for garment workers together with its 90 member companies. It is considered a frontrunner in working with brands on labor conditions in their supply chain. They represent over 120 of mostly up market and specialty clothing and sports brands with 15 production countries in Asia, Europe and Africa, including Bangladesh. Member products are sold in over 20,000 retail outlets in more than 80 countries around the world. However, the large buyers are largely absent from the membership, raising questions about how the key elements of the Fair Wear process could be further mainstreamed. The most commonly cited responses in the Impact Economy online survey to the question "which are in your opinion the most influential and promising standards?" were the Fair Wear Foundation, Fair Labor Association (FLA), Bluesign, Global Organic Textile Standard (GOTS), and the International Labour Organization (ILO) standards. Notwithstanding, nearly a quarter of respondents reported that they did not know or that they did not believe that any standard currently existed that was influential or promising in the global textile and garment industry. A widely agreed-upon indicator of what actually constitutes good practice does not yet exist.

Spotlight 9: Technology for the Transparency Arms Race

A major problem in the apparel industry is what insiders call "audit fraud" [13]. The term refers to the practice of inspecting an apparel factory that is not compliant with social and/or environmental requirements and to either detect no non-conformities (although they are apparent), or to note only easily fixable and relatively minor instances of non-compliance. For example, UK sustainable fashion journalist Lucy Siegle wrote about an auditor nicknamed "Eyewash" because he would ignore serious violations but would routinely check a factory's medical kit and note that it was lacking eyewash [14]. Documenting the actual working conditions in apparel factories is a difficult task—a respondent to the Impact Economy online survey commented that one of the worst practices of the industry was "copying [credentials] using the Internet and creating fake certificates via Photoshop" [15]. Gathering first-hand, real-time information on working conditions inside factories from workers is a potentially very powerful and disruptive way to increase supply chain transparency. For example, LaborVoices, a privately-held for-profit social enterprise based in California, United States of America (USA), began conducting a pilot in 2012 covering more than 20 factories with over 300 garment workers in Bangalore [16]. Using a secure communication platform to both send and receive messages, the software application was accessible to workers free of charge through their mobile phones and programmed in regional languages. The goal was to record local voices and conduct direct educational messaging on issues such as local labor laws, human rights, freedom of association and collective bargaining. This while protecting worker anonymity to hedge against reprisals. The sender dimension of the platform is a simple IT-based form of delivering information or training to enable workers to access relevant information with catalytic implications.

Aggregating information across factories, geographies, and industry segments through a bottom-up and mobile phone based mechanism can in principle generate a metrics-based stream of intelligence on corporate supply chains with clear action implications, provided a critical mass is reached. LaborVoices' business model consists of offering buyers direct and exclusive access to this intelligence on the factories they source from for an exclusive period of time. Exclusivity is time limited though, and the information is ultimately published more broadly—thus incentivizing proactive action or reactive damage control, depending on what the case may be.

Combining top-down efforts to enhance transparency via resourcing inspection capabilities with bottom-up mechanisms revealing the worker perspective, paired with social dialogue is the way forward. As defined by International Labour Organization (ILO), social dialogue "includes all types of negotiation, consultation and exchange of information between, or among, representatives of governments, employers and workers on issues of common interest" [17]. For example, in the "Worldwide Enhancement of Social Quality" project co-funded by German retailer Tchibo and the German Federal Ministry for Economic Cooperation and Development (BMZ), carried out by Deutsche Gesellschaft für Internationale Zusammenarbeit (GIZ), local trainers facilitated dialogue between managers and workers at 320 production facilities by the end of 2014, including in textile and garment factories [18]. Improved social dialogue paired with effective grievance mechanisms has the potential to improve social auditing significantly, and to enable more effective compliance with codes of conduct in the workplace [19].

While it would be naïve to assume that technology is a panacea for industry transformation, the overall march toward ubiquitous, low-cost, and real-time information gathering devices around the world is exciting. The mobile telephony revolution discussed in Part I opens up new possibilities to support other measures to achieve supply chain transformation with transparency around shop floor conditions and direct worker perspectives. To tip the balance in the de facto information technology arms race between the ever-lower cost of misinformation and prospect of greater transparency, a logical next step is to involve mobile phone operators in the solution design. The mobile telecom industry can be a catalyst for enabling higher performance in global value chains (and beyond). This especially in developing markets where the deployment of mobile telecommunications networks has surpassed traditional fixed-line technology [20]. Simple, inexpensive and convenient to use access to mobile networks is now widely available, even in remote areas; soon it will be possible for everyone and everything to be connected.

Research suggests that new mobile services could create potential annual livelihood benefits to workers globally of USD 7.7 billion by 2020, while also enabling a further USD 30.6 billion in benefits to organizations through improved productivity with an estimated 174 million in new service connections [21]. Accenture and Vodafone also argue that mobile communications can significantly improve working life and deliver commercial benefits to organizations in emerging and transitional economies. These include new job opportunities; improved working conditions; provision of timely, relevant training services; and secure delivery of wages using mobile money transfer solutions [21]. Furthermore, when assessing transformative solutions, we also need to consider the potential of cross-industry collaborations. This means reaching outside the customary set of stakeholders to come up with investable propositions that can drive profits and societal improvements on a larger scale.

5.3 Unlock the Hidden Treasures of Resource Productivity

Total resource productivity measures the production of goods with a given use of resources and externalities such as environmental impact. Utilizing as few resources as possible during a complete production process automatically involves cost efficiency, with decreased environmental impact. The money and externalities saved need to be evaluated against the investments required to raise total resource productivity. This is one of the core ideas to act on the dangers of an overshoot and collapse scenario originally identified in *Limits to Growth*. In the context of textile companies, water emissions can often be reduced by up to 50 % through the use of better inputs and technology, the optimization of production and the use of wastewater treatment technology. Energy use can be reduced by up to 40 % and chemicals by up to 20 % [22]. The implications of the industry's future shift to a circular economy are discussed below (Spotlight 10). But let us first look at an example.

MAS Holdings in Sri Lanka is a compelling illustration of the potential economic value that can be derived from upgrading [23]. In 2014, MAS Holdings was a leading clothing manufacturer in Sri Lanka, focused on lingerie and swimwear. Serious about sustainability, efficiency, and low environmental impact, the firm was also profitable. The MAS Intimates Thurulie factory, a 10,000-m^2 facility with roughly 1300 employees, was MAS Holdings' top plant, and also Leadership in Energy and Environmental Design (LEED) Platinum certified (LEED Rating systems are groups of requirements for projects that want to achieve LEED certification; each group is geared towards the unique needs of a project or building type). The factory cost USD 2.66 million to build, or 25 % more than conventional factories in Sri Lanka. The following details on the Thurulie factory provide a sense of what is possible in terms of environmental performance [23]:

- Passive design reduces heat loads; efficient evaporative cooling equipment maintains an indoor temperature of 27–29.5 °C (compared with 35–36 °C in an average Sri Lankan garment factory), while consuming only 25 % of the cooling energy of an average factory.
- Planted trees around the building ensure shading of the complex and grounds and keep the building an estimated 1–2 °C cooler.
- The thermal roof load, the largest contributor to heat gain and indoor discomfort in the tropics, is controlled by a combination of green roofs, photovoltaic roofs, and cool roofs. Green roofs cover 1757 m^2 of the building. They are installed on concrete decks over short span spaces in the administrative wing. Covered with turf and plants, the high thermal mass of this roof absorbs heat without transmitting it into the building. The cool roof is a lightweight metal roof assembly over the long span production halls. With a solar reflectivity index of 79, the roof reflects nearly 80 % of solar energy.
- The energy sources are photovoltaic and hydroelectric power, which are renewable and carbon neutral. The rooftop photovoltaic system with output of 25.6 kW covers 10 % of the plant's power needs. A small hydroelectric power plant connected to the public grid provides the other 90 % of the factory's power.
- Smart lighting is used as well. Work areas are illuminated with task lighting—high-efficiency T5 tubes and light-emitting diode (LED) lamps mounted on the sewing machines—focusing the correct amount of light at needlepoint. This system requires about half the normal number of light fixtures. Design features optimize natural light use.
- Sewing machines with direct-drive servo motors were chosen for energy efficiency, yielding a 15–30 % energy reduction in sewing.
- Every drop of water that enters the site is managed through rainwater catch systems. Consumption of potable water is about half that of comparable plants.
- Generated energy from solar and hydropower not used by the factory on holidays and weekends is sold back into the main power grid of Sri Lanka.

Spotlight 10: Circular Apparel—More Than a Crazy Idea?

In response to growing resource constraints, production in a variety of industries is now gradually shifting to a circular model, as discussed in Part I. This means that attention is increasingly focusing on the total resource productivity of the factors of production through materials, product and process innovation, and the avoidance of unnecessary waste (i.e., in packaging, as well as the reuse of waste and extension of the product lifespan). In apparel, the current industry model is similarly structured so that raw materials are utilized to manufacture goods that are then sold, used, and discarded as waste [23, p. 15]. The advent of "fast fashion" has led to even greater throughput of resources via apparel value chains. Researchers argue that fast moving consumer goods (FMCG) industries, including fashion, will move from the current model where 80 % of consumer goods are not recovered (and 18 % recovered for decomposition, and 2 % for reuse) to a model where non-recovery drops to 50 % in the near term [23, p. 86]. In this view, the shift to a circular economy results in a USD 595–705 billion cost savings opportunity, of which 10 % is in clothing [23, p. 56]. Some countries already have high collection rates of used clothing, for example the UK with 65 % [23].

We do not know how fast the transition to more circular business models will happen. But we do know that the next generation of sustainable apparel will have to do more than to comprehensively tackle working conditions and environmental footprints along the entire supply chain. It will have to shift from a linear economic model to a circular one. Two circular models seem to show particular promise in these current, early days of circularity: optimizing end-of-use by raising collection rates and recycling, and finding ways to have clothes circulate longer via collaborative consumption models [23, p. 54]. Radically greater resource efficiency in the production process could serve as a third possible circularity model. It is logical that once some leading players in the industry manage to integrate circularity in their business models, this will alter their cost structure, with the potential to create comparative advantage that forces other players to adapt as well. However, how soon such a development will take place is difficult to forecast. In short, circular apparel is not a crazy idea, but may take a while to become standard practice.

References

1. Daily Mail (2007) The battle of Primark: 3000 customers force their way into new store. http://www.dailymail.co.uk/news/article-446932/The-battle-Primark-3-000-customers-force-way-new-store.html. Accessed 1 Nov 2013
2. Chouinard Y et al (2011) The sustainable economy. Harvard Bus Rev. http://hbr.org/2011/10/the-sustainable-economy. Accessed 1 Nov 2013
3. Gereffi G, Frederick S (2010) The global apparel value chain: trade and the crisis. World Bank. http://econ.worldbank.org/external/default/main?pagePK=64165259&theSitePK=477894&piPK=64165421&menuPK=64166093&entityID=000158349_20100427111841. Accessed 7 Nov 2013
4. Nike Inc (2013) The Nike Responsibility register. http://www.nikeresponsibility.com/report/. Accessed 15 Nov 2013
5. Patagonia (2013) The Patagonia register. http://www.patagonia.com/us/environmentalism. Accessed 15 Nov 2013
6. H&M (2013) The HM register. http://about.hm.com/en/About/Sustainability.html. Accessed 15 Nov 2013

7. CSR and the Law (2010) The California Transparency in Supply Chains Act: new legislation requires disclosures on corporate efforts to eliminate slavery and human trafficking. http://www.csrandthelaw.com/2010/10/the-california-transparency-in-supply-chains-act-new-legislation-requires-disclosures-on-corporate-efforts-to-eliminate-slavery-and-human-trafficking. Accessed 14 Nov 2013
8. Office of the United States Trade Representative (2013) U.S. trade representative announces outcome of Generalized System of Preferences review. http://www.ustr.gov/about-us/press-office/press-releases/2013/june/gsp-review-outcome. Accessed 14 Nov 2013
9. Kim E (2013) Why penalizing Bangladesh isn't the answer. Businessweek. http://www.businessweek.com/articles/2013-06-28/why-penalizing-bangladesh-isn-t-the-answer. Accessed 15 Nov 2013
10. Office of the United States Trade Representative (2015) GSP review of Bangladesh recognizes progress, urges that more be done on worker safety and rights. https://ustr.gov/about-us/policy-offices/press-office/press-releases/2015/january/gsp-review-bangladesh-recognizes. Accessed 16 Jul 2015
11. Sustainable Apparel Coalition (2013) Apparel Coalition register. http://www.apparelcoalition.org. Accessed 14 Nov 2013
12. Fair Wear Foundation (2013) Fair Wear Foundation register. http://www.fairwear.org. Accessed 14 Nov 2013
13. Impact Economy (2013) Expert interview for sustainable apparel value chains project, Geneva
14. Siegle L (2011) To die for: is fashion wearing out the world? Fourth Estate
15. Impact Economy (2013) Online survey for sustainable apparel value chains project, Geneva
16. LaborVoices (2011) Bangalore garment worker pilot report. https://s3.amazonaws.com/lv-press-releases/LaborVoices+Bangalore+Pilot.pdf. Accessed 29 Nov 2013
17. ILO (2015) The ILO register. http://www.ilo.org/global/about-the-ilo/decent-work-agenda/social-dialogue/lang–en/index.htm. Accessed 16 Jul 2015
18. We Social Quality (2015) The We Social Quality register. http://we-socialquality.com/controller.aspx?n=1&l=2. Accessed 16 Jul 2015
19. Clean Clothes Campaign (2005) Looking for a quick fix: how weak social auditing is keeping workers in sweatshops. http://www.cleanclothes.org/resources/publications/05-quick-fix.pdf/view. Accessed 29 Nov 2013
20. Accenture and Vodafone (2011) Connected agriculture: the role of mobile in driving efficiency and sustainability in the food and agriculture value chain. http://www.accenture.com/SiteCollectionDocuments/PDF/Accenture-Connected-Agriculture.pdf. Accessed 5 Dec 2013
21. Accenture and Vodafone (2013) Connected worker: how mobile technology can improve working life in emerging economies. http://www.vodafone.com/content/dam/sustainability/pdfs/vodafone_connected_worker.pdf. Accessed 5 Dec 2013
22. Bluesign (2013) Water emission—improved protection of the natural circulation. http://www.bluesign.com/industry/bluesign-system/principles/water-emission#.UoTo1eJRHWg. Accessed 14 Nov 2013
23. MAS Holdings (2013) The MAS Holdings register. http://www.masholdings.com. Accessed 14 Nov 2013

Pull the Key Levers

6

6.1 Foster Total Resource Productivity and Transparency

A precondition for achieving higher total resource productivity is to understand what is possible. Transparency can help to lower access barriers to higher performance and achieve compliance with standards for producers. This can enable continuous improvement. On the consumer end, people love it when a seal or a statement about compliance with some standard simplifies complexity. Voluntary or "private" sustainability standards (hereinafter referred to in shorthand as "standards") enable value chain governance. This is especially useful in value chains that are long, cross-border, or unconventional. In principle, standards provide a mechanism that enables producers to learn about consumer requirements in global markets and adjust their setup and processes accordingly. Voluntary sustainability standards are one instrument among a broader set of regulatory options. Their adoption typically requires technical and financial support, as well as training. Such voluntary standards have one big advantage: they build on industry expertise, rather than classical regulation that tends to be several steps behind, as was apparent in the case of the financial services industry during the financial crisis. On the other hand, many standards are not particularly useful nor ambitious. The use of standards has been expanding rapidly since the 1990s. They are now impacting supply chain management and policy formulation in a number of industries. Additionally, a record of double-digit growth rates in sales of certified products (compared to those of conventional products) in a number of commodities is strengthening the business case for standards and certification (e.g. Organic, Fairtrade, Rainforest Alliance, and Utz Certified). Fairtrade continued to grow amid the financial crisis, increasing its global retail value by 15 % in 2009. Standards are also starting to influence public procurement on a significant scale. For example, the European Commission set the target of 50 % of 'green public procurement' by the end of 2013 for European Union member states [1].

M. Martin, *Building the Impact Economy*,
DOI 10.1007/978-3-319-25604-7_6

Notwithstanding, to raise the bar, taking measures to counteract confusion about the value add of particular standards is necessary. The experience from other industries has shown that corporate commitments to sustainable procurement and measurable tracking of goal achievement—the Mars and Cadbury commitment to the World Cocoa Foundation, or Unilever's engagement in sustainable sourcing of agricultural raw materials—can have a powerful impact, and be good business [2, 3]. In the case of the World Cocoa Foundation, Mars was the initiator but manufacturers started cooperating as they discovered that an increase of production volume (e.g. training of small scale farmers, better input factors and organization) was beneficiary to corporate business growth and helped to improve the reputation of the brand [2]. Mars is a family business though with a freedom to innovate. We need to ask ourselves how we can assist publicly listed corporations to find it easier to engage in such future-oriented sustainable business development strategies as well.

Standards and measurement methodologies in global apparel can be useful to gain transparency about the improvement potential a buyer has along different dimensions in the supply chain, such as health and safety, water, energy, waste or chemicals. Initiatives such as the SAC are useful because they put tools at the disposal of firms to engage efficiently in such optimization processes. However, being held accountable with external reporting, as in the case of Fair Wear and other standards, is also important: we need to make sure that action does not remain at the level of self-education and piloting, and that a better state of practice becomes the new normal over time.

6.2 Invest to Upgrade Industry Infrastructure

Supply chain upgrading will require significant investments. Many of these investments can ultimately deliver profits. Mobilizing capital is nevertheless challenging in the context of an industry where buyer-supplier relationships typically follow a transactional rather than a long-term mindset and investment targets are located in high-risk emerging markets with volatile order books and underdeveloped capital markets. Yet, considering the significant financial returns and social impact of value chain upgrading, developing or otherwise engaging in an impact-oriented investment strategy for the textile and garment industry is feasible. Allocating pilot capital can help to achieve a proof of concept.

Let's unpack the rationale. Providing investment capital for upgrading the physical infrastructure of factories provides a way to move beyond a reactive compliance approach. This is an important lever to drive industry change on the ground. The opportunity is conceptually straightforward: investment capital is scarce and expensive in developing markets. Given the generally low resource productivity in the industry, there is a win-win opportunity. It consists of improving social and environmental conditions while also enabling producers to save money and/or capture a higher margin by entering higher value added production.

Let's look at the opposite ends of the upgrading continuum first. Many factories work with extremely old—some might say, ancient—machinery. Even machinery that would merely be old in an advanced economy perspective is already an improvement. For example, one expert interviewee was very bullish about upgrading and pointed out how an investment of USD 80,000 in a pilot factory to replace machinery dating from the late 1960s with machinery from the late 1990s unlocked cost savings exceeding USD 500,000 over a subsequent twelve-month period [4]. We found similar economics looking at sample factories.

At the opposite end of the spectrum, consider very capital intensive upgrading and the case of Ananta Apparels Ltd (AAL). AAL is among the top ten of denim producers in Bangladesh and one of the country's leading apparel exporters [5]. In 2012, it had monthly production capacity of 1.8 million woven bottoms across three factories comprising 850,000 square feet of space and employing about 13,000 workers [6]. Major customers included leading retail brands such as H&M, GAP and Bestseller Group. With the financial support of the IFC, AAL embarked on an ambitious expansion plan that also aimed to introduce global best practices such as lean manufacturing, factory efficiency, worker productivity measures and achievement of energy efficiency standards. Relocation and construction are costly. Phase I primarily involved the relocation of AAL's existing casual woven bottoms factory to a new custom built facility in Kanchpur, south of Dhaka, thereby replacing its original factory in downtown Dhaka. Phase II involved further capacity expansion.

In a country where 90 % of buildings are not compliant with the building code, this is an opportunity to provide adequate physical infrastructure [7]. The project was estimated to cost approximately USD 24.4 million. The World Bank's private sector arm, the International Finance Corporation (IFC) committed to provide a financing package for up to USD 8 million. The expected social and environmental impacts include approximately 1500 new jobs and technology transfer. Moreover, new local supply chain linkages because the project is expected to generate incremental business for medium and small enterprises in the value chain, including accessory and package suppliers, service providers and logistics support. The government also stands to benefit once the project generates tax revenues [6].

Promising are moreover investment approaches that target a midpoint between the two opposite extremes discussed above. In a sample denim factory with approximately three million dollars in turnover, we found that a number of efficiency measures had a payback period under one year (Table 6.1), raising efficiency and improving worker health and safety.

If properly packaged, there is no logical reason why such investment opportunities, should not also be able to attract capital from impact-minded investors on a massive scale [9]. In fact, adopting an investment mindset is a core ingredient in the industry transformation formula. This is not bound to be a simple opportunity from the perspective of pure investment though. Each company has its own problems. Uncovering investment opportunities can be costly. One of the most critical challenges is local management bandwidth and transparency.

Table 6.1 Cost and resource savings in a sample factory

Measures implemented	Pre-implementation stage	Post-implementation stage	Installation cost
Measure	*Input volumes (monthly)*	*Potential savings*	*USD*
Eliminate water leaks in tubes and pipes	100,000,000 l	50 % water	60,450
Reuse cooling water from dyeing machines			
Reuse process water from rinsing			
Recover process water from bleaching			
Insulate pipes, valves, and flanges			
Reuse heat from drying operations	2,400,000 kWh	20 % electric power	8800
Improve steam management and insulate pipes, valves, and flanges	4,000,000 kg	5 % steam	0
Reuse cooling water from dying machines	25,000 kg	1.5 % indigo	5800
Reuse caustic soda	50,000 kg	2.5 % caustic soda	357,500
Total CAPEX			426,750
Monthly savings			43,000

Source author [8]

The time is nevertheless ripe for this next generation of investments in the textile and garment industry. Factories and supply chains are teeming with turnaround opportunities. Investors have not yet grasped them, especially where the fundamentals of products and processes that make heavy use of energy, scarce natural resources and low-skilled labor create a strong opportunity for financial returns that also result in social impact.

Moreover, the funding environment is changing. Facilitated by the work of the Accord and the Alliance, new financing facilities for upgrading are coming on stream. For example, in July 2015, the IFC announced making senior loans to finance factory repairs available for suppliers. In its first-ever deployment of long-term financing to Bangladeshi banks, the IFC aimed to provide USD 10 million each to five Bangladeshi banks. The banks are then to lend to local garment manufacturers. The loans finance upgrades as per the structural, electrical and fire safety corrective action plans (SEF CAPs) approved by Accord or Alliance [10]. Loans to SMEs are to be preferentially priced, carrying an interest subsidy financed by the UK Department for International Development (DFID).

6.3 Have a Fresh Go at Improving Working Conditions

The garment industry in many textile and garment producing countries outside of advanced economies grew extremely fast. Neither physical infrastructure nor human capital managed to keep up. While competitive private enterprise is the principal source of economic growth, wealth, and jobs, this does not automatically lead to an improvement in working conditions or sustainable livelihoods. Improving working conditions remains a top priority. Workers in the textile and garment industry, of whom 80 % are women, typically suffer from widespread insufficient health and safety conditions, violations of labor rights (e.g. wages, child and forced labor), inadequate housing, transport, healthcare and childcare, as well as gender discrimination. At the same time, there are issues with absenteeism and insufficient workforce productivity. Regardless of geography though, relentless short-term demand in the garment industry has meant a much greater focus on production rather than worker health and safety. One of the respondents to the Impact Economy online survey said her worst fear was that various emerging markets "will just continue in the same sweatshop pattern that we see now" [11].

Allocating resources to achieve substantial and wide-ranging improvement is a necessary part of any credible next generation solution leading to a more sustainable apparel industry. There are two main solution vectors. First, at the root of the problem is a widespread perception that decent work and competitive enterprises are in conflict with each other. Actions that aim to improve working conditions or safety are assumed to create additional cost burdens for enterprises when they can actually yield productivity returns. Research by the ILO and others suggests a potential mutually beneficial scenario. Improving workers' voice, empowerment, and skills through participatory management is a means to enhance working conditions and achieve efficiency improvements at the same time. Yet, the benefits that responsible practices generate for companies and workers alike are often systematically discounted, and not easy to demonstrate in the short term.

Better Work is a promising blueprint for illustrating the way forward on improving working conditions and productivity. A partnership between the ILO and the IFC, Better Work seeks to combine the ILO's expertise in labor standards with the IFC's acumen in private sector development. Better Work's training offering for all levels of a factory is instrumental in achieving a step change in working conditions. The training curriculum includes human resources management, Occupational Safety and Health (OSH), negotiation skills, training of trainers, supervisory skills training, preventing and addressing sexual harassment, and training on worker's rights and responsibilities. Better Work builds on the respective strengths of the ILO and the IFC: it highlights the need for effective interventions to combine expertise in social dialogue and the application of labor standards with a private-sector development and investment mindset. The vision is to enhance total resource productivity and environmental impact.

The main challenges of this approach, though, are its resource intensity and ability to embed it in factories for the long term. Newly established best practices could otherwise be abandoned once there is a change in management. One potential route is the prospect of combining the upgrading of factory infrastructure and working conditions with assuming (minority) ownership positions in factories. The rationale is that a one-time upgrade—in many cases subsidized by philanthropic, development finance, or public sector funding—needs to be locked in. One way to anchor such accountability is by bringing in investors who systematically pay attention to and track extra-financial "impact" variables at the manufacturing stage, such as working conditions or environmental footprint. This is good—provided the economics in what amounts to a highly competitive industry work out as well.

Next to a mutually enabling approach linking working conditions and productivity, grasping the role of gender is similarly important. In an industry where the workforce is composed predominantly of women, worker empowerment means women's empowerment. There are 4 million garment workers in Bangladesh alone (out of a total labor force of 77.6 million). They represent more than 90 % of the country's labor force in manufacturing [12]. Women constitute over 3.2 million or 80 % of this labor force [13]. This contrasts with female labor force participation of just under 30 % in the country overall [14]. What's more, working in apparel requires highly structured processes and precision, valuable skills in any post-garment career. The average household size in Bangladesh is 4.4 people. This means that the livelihood of more than 14 million people depends substantially on women working in the garment sector, or almost 9 % of the country's population [15].

Women play an important role in the modernization of gender roles and society. For example, unaccompanied women on the street are not aligned with traditional conservative values in Bangladesh. However, they are a logical social consequence of the expansion of the readymade garment sector, still a fairly recent phenomenon. In spite of their important economic contribution, however, women remain easy targets for exploitation and discrimination in factories. Their needs are not met on a systematic basis. Often unmarried and with poor education or training, women enter urban employment at the average age of 19 and with a comparative disadvantage in terms of pay, working conditions and the possibilities of promotion. They are typically concentrated in unskilled, low-paid, and often casual or informal work, including home-based work throughout the industry. Women's employment in the export-oriented garment industry has narrowed the gender gap in many spheres. Yet it remains visible with far too few women in supervisor and management positions. Most workers have little voice and influence in their workplaces. They are often denied the right to join a union or to organize. As a response, the Asia Floor Wage Alliance (AFWA), an international alliance of trade unions and labor rights activists who campaign for living wages for all garment workers, developed the Asia Floor Wage calculation to compute what constitutes a living wage across Asia [16].

The way the apparel work force is managed in many emerging market production locations is creating structural tensions. These play out in the factories, at home, and on the way to and from work. Factories often do not provide sanitary napkins or subsidize their cost even though menstrual-related absenteeism is common. Ignorance about health related rights such as maternity leave, factory child-care facilities, or nursing breaks is widespread. In Bangladesh, the combination of garment workers' long working hours and long waiting hours at state hospitals renders the hospitals practically inaccessible for the majority of factory workers. According to national legislation, factories with more than 300 workers should have a full time doctor for their workers. Most factories fail to fulfill this requirement and only provide a clinic room with very basic medicines.

Corroborating evidence for gender discrimination is solid. For example, of a total of 988 factory workers interviewed for another study, almost a third of the women reported sexual advances and being touched inappropriately at their workplace (297 and 290 respondents respectively) [17]. In addition to other forms of punishment such as being made to stand on tables, workers reported being subjected to threats [17]. Of those interviewed, almost half (484) reported being threatened with losing their jobs; over a third (333) reported threats of being sent to prison or threats of being forced to undress (328) [17]. There are large differences within the industry with respect to harassment and gender based violence. If we stick with the case of Bangladesh for a minute, we must also note that while verbal harassment and physical abuse seem to be widespread, rape is relatively rare and limited to smaller factories. More common is the use of highly sexualized vocabulary and body language as a means to discipline female workers. This creates a hostile, intimidating and sexually charged environment. Night work is associated with high risks of sexual assault or rape, with those working in the factories that are not in Export Processing Zones (EPZ) being the most vulnerable.

These observations offer a number of implications. Any upgrading strategy that does not aim to simply manage the problem needs to be absolutely realistic about the point of departure for female workers, and the contribution which skill training and productivity enhancements can make to both business and society. Behavioral change is hard and takes time. To foster the long-term sustainability of the industry, it needs to nevertheless contribute to strengthening the position of women beyond the factory gates. This means that one needs to assess the extent to which programs that are focused on working conditions and education are actually contributing to solving gender-related issues. Actors should also be prepared to take additional action where needed, and understand the dynamics associated with larger shifts that touch political empowerment and the societal fabric.

This will require a departure from the currently prevalent modus operandi. Several expert interviewees pointed out that interventions in the industry that address working conditions often look at the problem through a "labor" prism. This opposed to fully grasping the gender dynamic and the limitations that result from traditional trade union structures that do not fully represent women's voices. To achieve supply chain transformation, a new level of ambition is required for improving working conditions. Productivity and gender are both key. Successful implementation will require transcending longstanding adversarial NGO-private sector relations to enable a quality rollout in a sufficient number of factories to set in motion a change process on a much larger scale. Moreover, improving working conditions cannot be disassociated from upgrading production inputs, processes and infrastructure.

6.4 Learn from Leading Players

Breaking out of the mold of constantly reinventing the wheel is key to achieving scale. There is a lot to be learned from best practices in social and environmental performance and value added creation. A meaningful way of keeping the cost of learning down is to get involved in pilot initiatives that draw from the lessons and practices of others. Fire and building safety in particular is an area where work is underway to improve the state of affairs. To address the issue of fire and safety post-Rana Plaza, over 190 brands signed a legally binding Accord, which was initiated by a group of international labor unions (e.g. Industriall, Global Union and the UNI Global Union) in partnership with leading NGOs (e.g. Clean Clothes Campaign and the Workers' Rights Consortium), and funded by European brands [18, 19]. US retailers and apparel brands subsequently formed the Alliance for Bangladesh Worker Safety to tackle the same issues in a non-legally binding structure more adapted to the realities of the US legal system [20]. Both the Accord and the Alliance were direct responses to the Rana Plaza accident, and their inspection reports and corrective action plans provide a structured framework to render upgrading across the board possible. The fire at Dignity Textile Mills Limited in Gazipur on May 31, 2015, a factory producing for Accord signatory companies, serves to showcase both the progress and the work still ahead [21]. The Accord had inspected the building about a year earlier, requiring installation of an automatic sprinkler system throughout the building. The fire started while many of the workers were outside the building having their lunch, with still more than 1000 workers in the building when the fire alarm sounded. The workers left the factory in an orderly manner, avoiding casualties. However, the fire kept burning for over 24 h, destroying much of the factory's steel structure. While many of the issues related to electrical safety had been corrected by the time of the Accord's follow-up inspection two months before the fire, the factory had not yet installed the sprinkler system. As the Accord pointed out in its press statement, "Had a properly designed, installed and maintained fire sprinkler system been in place, those 3000 garment workers employed at the factory would be at work again" [21].

What is now needed is movement toward a next generation of initiatives that build on what has been achieved, but also take a broader view of enhancing social, environmental and economic performance. For example, the Responsible and Accountable Garment Sector (RAGS) Challenge Fund, supported by UKaid from the Department for International Development (DFID), and its grantees provide valuable lessons to explore for upgrading activities in the field [22]. RAGS was established as a grant fund with British Pound Sterling (GBP) 3 million to support eligible projects over a three-year period (2010–2013). It supported the work of a wide range of organizations in the garment supply chain, including large and small businesses, labor, fair trade and other non-governmental organizations (NGO). The common denominator was the effort to catalyze better working conditions for workers in readymade garment production industries in low-income countries in

Asia and Sub-Saharan Africa, which supply the UK market. Even so, a closer look also reveals the challenges any industry transformation model has to address. RAGS was a discrete project without a sustainable long-term financing model; this is simply the way the development industry normally works. However, even if RAGS was not financially independent on standalone terms, financial and technical sustainability plans were built into the funding approval process, and a number of the supported projects had post-RAGS measures for financial continuation from their own and/or other sources of funding [4]. While RAGS has brought about an admirable and positive impact for a large number of workers, the durability of these achievements is less clear once funding ends. Moreover, reducing worker turnover at the factory level from 9 % a month to 5 % a month is considered a major success in the industry. The unavoidable fact is that, on average, every worker trained at considerable expense will most likely work in a new institutional environment within 20 months.

RAGS played an important part in the UK Sustainable Clothing Roadmap—a voluntary clothing industry initiative established and managed by the Department for Environment, Food and Rural Affairs (Defra), but implemented by the Waste and Resources Action Programme (WRAP), Defra's delivery body, to improve the environmental and ethical performance of clothing [23]. The UK Sustainable Clothing Roadmap evolved into the Sustainable Clothing Action Plan (SCAP), with a focus on carbon, water and waste across the clothing lifecycle, leaving issues relating to labor and chemistry to other players [4]. The key questions remain however: (1) how to scale the types of interventions RAGS is targeting so they can achieve critical mass (including managing their cost), and (2) how to make them durable in time.

The next generation of industry transformation efforts will need to make a more explicit connection between how higher social and environmental performance can actually drive revenue creation via increased resource productivity and savings. The key question is how to anchor achievements long after a single upgrading project has ended, and workers have moved on to new employers. This is particularly true for addressing the human capital gap. Without technical skill, the ability to enter higher value added production and raise productivity hits bottlenecks. Without management skill, adversarial labor relations lock up all parties into unproductive patterns of behavior and lower efficiency. This undercuts the ability to pay higher wages. The shift of the global apparel industry to the developing world has historically meant a tremendous loss of knowledge and skill. However, the vision of a sustainable global apparel industry will require greater skill at all levels—from selecting and deploying chemicals at the beginning of the pipe to in-country pre- and post-production services such as design and quality testing. This requires human capital on a massive scale, as well as leveraging the power of information technology.

References

1. European Commission (2015) EC Europa register. http://ec.europa.eu/environment/gpp/faq_en.htm#general1. Accessed 15 Jul 2015
2. World Cocoa Foundation (2013) The World Cocoa Foundation register. http://worldcocoafoundation.org. Accessed 2 Nov 2013
3. Unilever (2013) The Unilever register. http://www.unilever.com/aboutus/supplier/sustainablesourcing/index.aspx. Accessed 2 Nov 2013
4. Impact Economy (2013) Expert interview for sustainable apparel value chains project, Geneva
5. CDC Group (2013) Ananta Apparels, Bangladesh. http://www.cdcgroup.com/Global/Case%20Studies/Documents/Final%20CDC-4Page_Ananta_210513.pdf. Accessed 2 Nov 2013
6. International Finance Corporation (2012) Ananta Apparels Limited: environmental and social review. http://ifcext.ifc.org/ifcext/spiwebsite1.nsf/651aeb16abd09c1f8525797d006976ba/6ae5d89f3a1b407785257ab4006d32a6?opendocument. Accessed 10 May 2015
7. The Economist (2013) Rags in the ruins. http://www.economist.com/news/asia/21577124-tragedy-shows-need-radical-improvement-building-standards-rags-ruins. Accessed 4 May 2013
8. Martin M (2015) Financial model for Apparel Innovation Consortium project. Impact Economy
9. Martin M (2013) Status of the social impact investing market. UK Cabinet Office. https://www.gov.uk/government/publications/status-of-the-social-impact-investing-market. Accessed 2 Nov 2013
10. IFC (2015) IFC teams up with leading banks, buyers, to improve safety in Bangladesh garment factories. http://ifcext.ifc.org/IFCExt/pressroom/IFCPressRoom.nsf/0/511BAE10B2E5B2C785257E7B006B6CD0. Accessed 16 Jul 2015
11. Impact Economy (2013) Online survey for sustainable apparel value chains project, Geneva
12. World Bank (2014) Bangladesh data sheet: Bangladesh total labor force. http://data.worldbank.org/indicator/SL.TLF.TOTL.IN/countries. Accessed 15 May 2015
13. Bangladesh Bureau of Statistics (2013) Bangladesh data sheet. http://www.bbs.gov.bd/WebTestApplication/userfiles/Image/SubjectMatterDataIndex/datasheet.xls. Accessed 1 Dec 2013
14. ILO (2013) Decent work country profile Bangladesh. http://www.ilo.org/wcmsp5/groups/public/—dgreports/—integration/documents/publication/wcms_216901.pdf. Accessed 4 Dec 2013
15. Bangladesh Bureau of Statistics (2013) Average household size by sex of head of household and residence. http://www.bbs.gov.bd/RptHIES_2_1.aspx?page=/PageReportLists.aspx?PARENTKEY=73. Accessed 3 Dec 2013
16. Merk J (2009) A decent wage across borders. http://www.asiafloorwage.org/documents/Stitching%20a%20Decent%20Wage%20Across%20Borders.pdf. Accessed 7 Nov 2013
17. War on Want (2011) Stitched up: woman workers in the Bangladeshi garment sector. http://www.waronwant.org/attachments/Stitched%20Up.pdf. Accessed 7 Nov 2013
18. Industriall (2013) Bangladesh Accord on Fire and Building Safety released. http://www.industriall-union.org/bangladesh-accord-on-fire-and-building-safety-released. Accessed 7 Nov 2013
19. Clean Clothes Campaign (2013) We made it!—global breakthrough as retail brands sign up to Bangladesh factory safety deal. http://www.cleanclothes.org/news/press-releases/2013/05/16/global-breakthrough-as-retail-brands-sign-up-to-bangladesh-factory-safety-deal. Accessed 7 Nov 2013
20. Alliance for Bangladesh Worker Safety (2013) The Bangladesh Worker Safety register. http://www.bangladeshworkersafety.org/. Accessed 7 Nov 2013
21. Bangladesh Accord (2015) Accord statement on fire at Dignity Textile Mills Limited, Natun Bazaar, Sreepur, Gazipur. http://bangladeshaccord.org/2015/06/accord-statement-on-fire-at-dignity-textile-mills-limited-natun-bazaar-sreepur-gazipur. Accessed 16 Jul 2015

22. Department for International Development (2013) Responsible and accountable garment sector challenge fund. https://www.gov.uk/responsible-and-accountable-garment-sector-challenge-fund. Accessed 7 Nov 2013
23. Department for Environment Food and Rural Affairs (2011) Sustainable clothing roadmap. https://www.gov.uk/government/uploads/system/uploads/attachment_data/file/69299/pb13461-clothing-actionplan-110518.pdf. Accessed 29 Nov 2013

How to Chart a Path for Supply Chain Transformation

7

7.1 The Future Can Be Dynamic and Positive

In the 1860s, an English discoverer named C.T. Beke proposed to construct a 225-mile railroad to open up trade with the southernmost territories ruled by the Pasha of Egypt, who nominally reported to the Ottoman Empire [1]. His motivation was to provide ready access to source from the cotton fields of Ethiopia, connecting the coast of the Red Sea with the Upper Nile Valley. Nearly 150 years later, the Ethiopian Government, through its Growth and Transformation Plan (GTP) was engaging in massive industrial and infrastructure projects; a 1500 mile-long standard gauge rail network to help overcome the landlocked country's infrastructure limitations and render a national trade logistics strategy viable being among them.

The first textile and garment factories in the country were established back in 1939. The textile and garment industry is among the sectors especially incentivized by the government, with tax holidays and capital goods imported free of duty. Due to its very significant cotton production potential in higher altitudes (between 1000 and 1400 m)—provided that irrigation compensates for lack of sufficient rainfall—Ethiopia has the ability to establish a complete textile and garment supply chain, which is unlike some other major sourcing hubs that need to import most of the inputs. Further, the country's proximity to European markets provides a major locational advantage, provided Ethiopia can develop more efficient logistics, and possibly additional trade corridors, to reduce reliance on the port of Djibouti to handle exports. Today, Ethiopia's government is keen to see the apparel industry reach exports of one billion dollars by 2016, and is actively inviting other countries and major industry players to invest. With textile and apparel exports valued at USD 111 million during the 2013–14 fiscal year, achieving such growth may take longer, but high ambition and the underlying potential are observable already [2]. As the Economist Intelligence Unit (EIU) suggests, next to infrastructure investments under way and a young population with a median age of only 17.6 years, the cost of inputs such as land, water and power is low. Ethiopia has a land area suitable for cotton plantations rivalling that of Pakistan, the world's fourth-largest producer [3].

M. Martin, *Building the Impact Economy*,
DOI 10.1007/978-3-319-25604-7_7

The future can be dynamic and positive. Foreign companies have started entering Ethiopia. For example, Turkey-based Akber is constructing the country's largest textile plant worth USD 175 million, the India-based ShriVallabh Pittie Group (SVP) Group is setting up a USD 550 million spinning mill, and the Chinese textile company Zhejiang Jinda Flax Llc aims to develop an entire textile industrial zone [1]. Yet all of these developments are taking place against the background of an overall industry that is actually less sustainable today than it has ever been. As it gears up, a key question is how Ethiopia could steer clear of the social and environmental problems that afflict other sourcing countries.

Moreover, disruption, in its many forms, is on its way, not just spatially, but also in terms of technology. In this spirit, one respondent to the Impact Economy online survey commented: "Within a 10–20 year period I think the industry will be changed severely by higher customer awareness on labor standards and environmental issues, production techniques such as robotics, 3-D printing and new materials, and increases in costs due to higher raw material prices" [4]. This has every potential to be positive. But disruption also brings with it any number of foreseen and unforeseen perils (as an aside, among other things, we will certainly have to think through whether robots will also have to pay taxes as they become more numerous and intelligent, just like humans need to).

Right now though, we have more urgent issues to attend to. Many consumers may either not care or erroneously assume that luxury means quality all along the supply chain, but the time to step up the game for brands ranging from value to luxury is nevertheless now. From a company perspective, the case rests mainly on total resource productivity and managing reputational risk. Views such as that of an online survey respondent who argued "Social compliance has become a threshold requirement to doing business; you simply cannot compete without it. Therefore, there is no trade-off—social compliance is necessary to success" are not yet the new normal, but can become so in due course [4].

Upgrading the global textile and garment industry so that supply chains are transparent, working conditions adequate, and environmental footprints are optimized will require doing a lot of homework and building a coalition for change. The verdict on Bangladesh is still out. But industry consolidation and repositioning is already underway in places like China and Japan. We do not know how the world will look in 2050, but we do know that there will be demand for apparel products as long as there are people. 9 billion people will have to clothe themselves, not to mention look and feel good while doing it.

However, from the stampede of shoppers on Oxford Street, to the rioting of protesters in Bangladesh, the apparel industry of today is not yet up to the task of meeting the demands of this future in a sustainable and profitable fashion. What is the alternative? A respondent echoed this sentiment in dramatic fashion when he indicated that his worst fear for the future of the international textile and garment industry is that it will "become the breeding ground for modern slavery" [4]. Spotlight 11 below provides a glimpse at the dynamics in Myanmar, Asia's most recent garment frontier.

Spotlight 11: Myanmar—Back- or Forward for Asia's Frontier?

The reform movement occurring in Myanmar (Burma) is serving to open up a new level of opportunity for a renewed sourcing location in the apparel industry—with the potential to do better than other apparel production clusters that have led to poor social and environmental performance and low value added. Myanmar assumed the Association of Southeast Asian Nations (ASEAN) Chairmanship in 2014, and faces an important national election in the fall of 2015 [5]. GDP growth is forecast to average 8.2 % a year from 2015/16 to 2018/19; investment is expected to expand by 14.3 % a year on average (up from 12 % in 2000/01) [6].

In terms of overall size, Myanmar's readymade garment industry is currently small, projected to earn about USD 2 billion in 2015/2016. This up from USD 770 million in 2011 [7]. According to Dr. Aung Win, vice chairman of the Myanmar Garment Manufacturers Association (MGMA), roughly 350 garment factories currently operate in the country, which is tiny when compared to major sourcing locations in South Asia [8]. Apparel manufacturers cut and sew woven or knitted fabric or knit apparel directly from yarn. The cut-and-sew classification includes a diverse range of establishments making full lines of ready-to-wear and custom apparel. Apparel manufacturers can be contractors, performing cutting or sewing operations on materials owned by others, or jobbers and tailors who manufacture custom garments for individual clients. Firms can purchase textiles from another establishment or make the textile components in-house. In Myanmar, factories are operating based on a Cut, Make, and Trim (CMT) approach. Over time, Myanmar aims to upgrade its production systems from a contract manufacturing (more than 90 % of national factories) to Original Equipment Manufacturing (OEM) and Original Brand Name Manufacturing (OBM) in order to increase value added. This shift will need to be supported by an adapted legislative framework. Estimates for export potential are about USD 5–6 billion even without major upgrading [8].

Myanmar is roughly the size of Germany and Italy combined. With 61 million inhabitants, as well as a migrant population, the country has the potential for significant market expansion [9]. It is presently going through an opening up to the world after decades of essentially closing itself off to foreign interests (and vice versa). American sanctions initiated with an arms embargo in 1993 and widened in 2003 to include all new investment, as well as a ban on imports from Myanmar, have had the expectable negative impact on the export capacities of the garment manufacturing industry [10]. Current leading investors in the national garment industry are China, Japan and Korea, sometimes through local strawmen. Most recently, the Myanmar Investment Commission (MIC) has begun to approve foreign investment in the country's garment manufacturing industry [11]. This has generated major interest from international investors.

Beyond foreign direct investment (FDI) policy changes, other critical features fueling the growth of the apparel industry in Myanmar relate to human capacity and installed infrastructure. The country is logically positioned to fulfill the basic equipment manufacturing needs that other markets can no longer easily or affordably satisfy. The threat of a growth path that simply leads to a production cluster with thousands of sweatshops is real. This would recreate the abuses and controversies associated with the apparel industry in neighboring Bangladesh. A respondent to the Impact Economy online survey commented that one of his worst fears is a "transfer of industry to Myanmar with no local NGOs or unions, [and] exploitation of workers" [4].

As the reform path continues, the garment industry has the potential to grow well beyond six billion dollars in annual exports—provided upgrading is successful and issues such as compliance with basic worker issues and factory safety standards, standardized payroll, and minimum wages can be addressed productively. This would also give a major boost to national development. Sustainable production is important for social peace, attracting foreign investment and the awarding of duty free access to the EU—all milestones to usher in an era of substantial sector growth [12]. There are indications that some very innovative thinking how to reconcile competitiveness, environmental and social performance is now advancing in frontier markets such as Myanmar. Consider that in 2015, two NGOs committed to better working conditions, the Ethical Trading Initiative and the Fair Labor Association, won the backing of 17 major international manufacturers—including adidas, Gap, Tesco, H&M and Patagonia—in their advocacy efforts to achieve a higher minimum wage for workers in the country's garment sector [13–15]. The possibility of moving forward is real.

7.2 How to Design Complementary Models that Can Scale

If all T-shirts were produced using the latest available technology, this would save 26 % of energy. There is machinery on the market in wet processing that saves EUR 160,000 in annual running cost, compared to a cost of purchase of EUR 250,000 [16]. Leading solutions exist. Yet getting them out there in scale is the challenge. In the short term, and as the examples of Ethiopia and Myanmar above indicate, simply pursuing the next country frontier is still the modus operandi for many. But the Rana Plaza disaster and other industrial accidents illustrate that an unrelenting and unforgiving pursuit of growth is risky. Fortunately, enormous value-creation opportunities are waiting to be tapped—opportunities that will provide the economic and social basis for changing business as usual.

Granted, the puzzle laid before us may be complicated, pieces are missing, and the stakes are high. But this is just like in the other fields explored throughout this book. If we start now, we can begin to link what works, learn from what doesn't, and wager for a future that is both profitable and sustainable. Looking at one specific industry, the preceding chapters argued that systemic upgrading is possible if key stakeholders are bold, cooperate, and dare to act.

Change often starts as the lone creative act of a leader or innovator; but no transformation can be achieved without building a grander coalition for change. To be successful, it is not enough for the envisioned change to be "better" than other industry solutions; it must also attract a critical mass of backers, representing the different types of stakeholders discussed earlier. The collaboration between a development agency and a leading nonprofit illustrates the value the development industry can bring to the process of supply chain transformation, rather than overburdening local businesses with handling the challenge completely on their own.

To facilitate such a transition toward upgrading and impact-oriented management of productive assets in the main sourcing countries, the development industry could make a contribution by incentivizing stakeholders to embrace next generation models. Conceivable instruments include co-deploying investment capital alongside grants, and thereby facilitate capital mobilization by other players, or providing funding incentives for internalizing externalities such as water use, which often do not carry a cash cost in production locations. To move forward on supply chain upgrading, it is key to be pragmatic and have a toolbox of solutions, some financial, some technical, some organizational. Some tools are best suited to financing larger "demonstration" projects. Other tools fit the bulk of the (smaller) factories especially well.

To provide a blueprint for Part II's relevant *how-to* question—how to make all of this happen when the investment sizes are too small for a classical investment approach—, a consortium approach is well suited. For example, to render the insights gathered from the post-Rana Plaza study "Creating Sustainable Apparel Value Chains" actionable, we developed the model of an Apparel Innovation Consortium (AIC) [17]. The AIC comprises a number of stakeholders who pool resources in order to create a sustainable and viable textile and garment industry. It aims to improve social and environmental conditions in textile and garment manufacturing by raising the productivity of its resources. To this effect, it uses a combination of debt funding, grant funding, efficient procurement, and measures which can be standardized to some extent. These are replicable interventions, which are needed in a large number of textile and garment factories, such as re–using heat from drying operations or improving steam management, paired with introducing effective social dialogue. Replicating such measures allows for economies of scale and scope in applying them. To make a step up on the ladder toward systemic change, the AIC is designed to target the "lowest hanging fruits"—by providing solutions to the problems, which provide the greatest positive results for a given amount of resources deployed. Rather than fixing everything at once, only to remain stuck at the level of a handful of factories, the approach focuses on the introduction of a self-scaling investment logic.

Such a consortium approach entails a broader perspective than a private equity firm could have. A private equity firm, which is considering investing in garment factories in emerging markets and creating value by modernizing them would need to earn a cost of capital of 20 % or more on its investments. To achieve such a return, one must focus on larger factories. If such a larger-scale approach were to be used alone, the many smaller factories, which make up the industry, would not be included. The AIC approach, on the other hand, uses financial engineering and taps different pools of capital to act in the set of factories that comprise the bulk of a sourcing country's industrial garment assets. It is on purpose complementary in that it focuses on certain basic improvements, which work in most factories and which can yield impressive results in terms of environmental and social improvements, as well as financial savings. For example, if leaks are eliminated from tubes carrying water for a company's dyeing process, or if the company's heat and water are reused, the company will see a fast financial return on those small changes, in many cases paying for themselves in under one year. Improvements in productivity and in the use of resources free up money to spend on labor conditions and investments in the factory.

7.3 Take up Fashion's Ambition Level by More Than a Notch

If we are serious about avoiding overshoot and collapse, we must find a more effective way to (re)use our resources and handle human and environmental impacts. The global apparel industry has experienced many new ideas and technologies over the centuries. As we have moved from the agrarian to the postindustrial age, it has both changed and been influenced by our tastes and aspirations—serving fundamental human needs in the process, just like many other industries that have become global over the past decades. Today, though, changing and often less predictable seasonal consumer demand, a variety of market trends, short product life cycles, and low barriers to entry have all made the industry much more competitive [18]. At the same time, the demand for socially and environmentally compliant apparel is gradually rising.

Solutions to social issues are difficult, though social enterprise and impact investing have ushered in an era of innovation in social service provision and longer-term minded responsible investment capital. And as the information revolution fosters ever-greater transparency, sustainable forms of consumption will catch on eventually. The reader will recall that in the case of CIV discussed in Part I, I argued that forward-looking companies need a methodology to engage in holistic business innovation to stay profitable and relevant as this century's sustainability imperative unfolds. At the company level, this will drive profits and impact; at the aggregate level, it will enable mastering our collective sustainability and wealth creation challenges. Looking at the textile and garment industry, it is similarly essential to develop solution approaches that can be used in any manufacturing location. The problem is too big and pervasive to keep ourselves busy with putting bandaids on symptoms.

Fortunately, our answers to the *how-to* questions can be specific: taking textile and garment supply and value chains to the next level is possible, and the measures are known. This holds lessons for other labor-intensive long supply chains as well. Favorable trends in green growth and total resource productivity, the rise of the LOHAS consumer, and growing demand for affordable products and services at the Base of the Pyramid are all creating an environment that is priming global supply chains to help drive inclusive growth.

The goal now is to overcome inertia because of the lack of a self-evident transition path. A bolder path for industries needs to be charted. Hence this book's relentless focus on the *how-to* questions. The Part III looks at energy, the mother of all markets. This to understand what needs to happen to lay the foundation for a comprehensive sustainable transformation of the global economy: a successful energy transition that can provide cheap, reliable, carbon-neutral energy, and thereby enable the conversion processes characteristic of the largely closed-loop economy, which we will ultimately need to build.

References

1. Martin M (2014) Grasping sustainability: a world-class textile and garment industry can become Ethiopia's finest asset. In: Addis Standard. http://addisstandard.com/grasping-sustainability-a-world-class-textile-and-garment-industry-can-become-ethiopias-finest-asset/. Accessed 16 Jul 2015
2. Alemayehu SA (2015) Ethiopia: reauthorization of Agoa—opportunity to boost Ethiopia's exports. http://allafrica.com/stories/201507141256.html. Accessed 17 Jul 2015
3. Mosavi S (2014) Made in Ethiopia: the rise of the textile industry. http://www.economistinsights.com/opinion/made-ethiopia-rise-textile-industry. Accessed 16 Jul 2015
4. Impact Economy (2013) Online survey for sustainable apparel value chains project, Geneva
5. Bangkok Post (2013) Don't ignore Myanmar politics, Suu Kyi Tells EU. http://www.bangkokpost.com/news/local/379895/don-t-ignore-myanmar-politics-suu-kyi-tells-eu. Accessed 15 Nov 2013
6. Hamstra J (2013) Country report Myanmar. Economist Intelligence Unit. http://www.eiu.com/default.aspx. Accessed 7 Nov 2013
7. The Daily Observer (2015) Myanmar to expand RMG sector for growth. http://www.observerbd.com/2015/04/08/82441.php. Accessed 15 May 2015
8. Impact Economy (2013) Expert interview for sustainable apparel value chains project, Geneva
9. Asian Development Bank (2012) Myanmar facts sheet. http://www.adb.org/sites/default/files/pub/2013/MYA.pdf. Accessed 7 Nov 2013
10. BBC (2009) Overview of Burma sanctions. http://news.bbc.co.uk/2/hi/asia-pacific/8195956.stm. Accessed 22 Nov 2013
11. Fibre2Fashion (2013) Myanmar approves new foreign investments in garment sector. http://www.fibre2fashion.com/news/apparel-news/newsdetails.aspx?news_id=152332. Accessed 2 Nov 2013
12. David Birnbaum's Blog (2013) The Myanmar garment industry: a very short study. http://www.birnbaumgarment.com/2013/02/28/the-myanmar-garment-industry-a-very-short-study/. Accessed 2 Nov 2013
13. Barron, L (2015) High-profile foreign brands back minimum wage for all. http://www.mmtimes.com/index.php/national-news/15629-high-profile-foreign-brands-back-minimum-wage-for-all.html. Accessed 23 Jul 2015
14. Democratic Voice of Burma (2015) Labour watchdogs support Myanmar minimum wage. In: Bangkok Post. http://www.bangkokpost.com/news/asean/624956/international-watchdogs-support-myanmar-minimum-wage. Accesssed 16 Jul 2015
15. DVB (2015) International brands support minimum wage, say watchdogs. http://www.dvb.no/news/international-brands-support-minimum-wage-say-watchdogs-burma-myanmar/54817. Accessed 16 Jul 2015
16. Fassi M (2015) Eco-co approach for a sustainable yarn dyeing plant. In: PowerPoint presented at the 4th Bluesign Conference, St. Gallen, Switzerland, 13–14 July 2015. http://www.bluesign.com/sites/fourth-bluesign-conference/downloads#.VbeOeLqyM3s. Accessed 28 Jul 2015

17. Textiles Intelligence (2014) Talking strategy: Maximilian Martin of Impact Economy discusses how competitiveness in the apparel industry can be improved by investing in sustainability and productivity. Glob Apparel Markets 28(4):4–18
18. Islam M et al (2013) Textile industries in Bangladesh and challenges of growth. Res J Eng Sci. http://www.isca.in/IJES/Archive/v2i2/7.ISCA-RJEngS-2013-002.pdf. Accessed 25 Oct 2013

Part III
Fixing Energy, the Mother of All Markets

It is not a dream, it is a simple feat of scientific electrical engineering, only expensive—blind, faint-hearted, doubting world! [...] Humanity is not yet sufficiently advanced to be willingly led by the discoverer's keen searching sense. But who knows? Perhaps it is better in this present world of ours that a revolutionary idea or invention instead of being helped and patted, be hampered and ill-treated in its adolescence—by want of means, by selfish interest, pedantry, stupidity and ignorance; that it be attacked and stifled; that it pass through bitter trials and tribulations, through the strife of commercial existence. So do we get our light. So all that was great in the past was ridiculed, condemned, combatted, suppressed—only to emerge all the more powerfully, all the more triumphantly from the struggle.

Nikola Tesla[1]

[1]Good Reads (2015) Problem of increasing human energy quotes. http://www.goodreads.com/work/quotes/4763-problem-of-increasing-human-energy. Accessed 21 July 2015.

No Disruption, No Exponential Progress

8

8.1 Solex Needed: We Cannot Afford to Get Energy Wrong

If we strip away the alarmism about a world polluted beyond redemption and a global civilization in catastrophic decline, the transformational new insight the simulation model in *Limits to Growth* offered when first published in 1972 were the several pathways how a population can adjust to its habitat's ultimate carrying capacity, in an environment that is characterized by finite resources. This either by reducing the growth rate or by overshooting the environmental limit, to only then converge back in some fashion on the carrying capacity level [1]. Since the overshoot could itself reduce that environmental limit permanently, the new equilibrium might well be set at a lower level—just think about the trees, or rather the lack thereof, in South Italy today, more than two thousand years after the Romans had cut most trees to build the fleets that would enable their victory against Carthage. As a result, a large public first realized that the motorway of economic growth would end eventually—and perhaps abruptly.

In today's resource hungry economy and with a growing population, a logical way forward is to reduce the resource intensity. Ideally, one would run the economy in a closed loop, where resources are recycled and reused. This requires massive amounts of energy however. The ability to provide energy from renewable resources becomes itself a critical constraint. There is a silver lining on the horizon though: a lot is possible that until recently would have sounded like science fiction. For example, what if mobile phones could be charged superfast in 30 s, and electric cars in three minutes? This may become reality in 2017: similar to proteins used by body builders to grow bigger faster, StoreDot—a start-up born from the nanotechnology department at Tel Aviv University (Israel)—used biological semiconductors made from naturally occurring organic compounds known as peptides—short chains of amino acids, which are the building blocks of proteins—to accomplish this. In 2017, StoreDot plans to release a fast-charging battery capable of replenishing an electric car to full in just 3 min [2].

M. Martin, *Building the Impact Economy*,
DOI 10.1007/978-3-319-25604-7_8

Part III focuses on the role of energy in building the Impact Economy. Over the past years, the topic of clean energy has kept moving up on the international agenda. Many major oil and gas companies, who are at home in the fossil fuel business, are currently reviewing their policies on climate change. On June 1, 2015, six of Europe's largest oil and gas companies even banded together for the first time, asking the UN to let them help devise a plan to stop global warming, in preparation of the 2015 Paris talks on climate [3]. Some are considering renewing their previously unsuccessful participation in the renewables industry dating back to a decade ago [4].

To put things in perspective, some historical context is useful. Compared to the end of the 18th century—when the previous renewable and "low" energy era ended —, the sheer amount of energy consumed has increased by several orders of magnitude. At the time of the French Revolution in 1789, Europe was mainly powered by animals—14 million horses and 24 million cattle with a total power of 7.5 billion Watts, or equivalent to the power of about 100,000 mid-size cars today [5]. 600,000 water mills and firewood complemented the largely renewable energy mix two hundred years ago.

The subsequent fossil fuel revolution rendered economic development and modernity, as we know it, possible. When this happy narrative was disrupted by *Limits to Growth*, intense and often emotional debate on what will happen, with doomsday scenarios and projections of bright futures put forward alike, became a stock of the public debate [6]. Disaster has not yet struck. But today, scenario planning is well established.

Rather than projecting the future, throughout this book our interest is therefore in what has not been as well explored: the '*how-to*' question of how we can practically *get* to our preferred future. Just like Chinese leader Deng Xiaoping's famous maxim that "It doesn't matter whether a cat is white or black, as long as it catches mice," it does not matter if solutions are "unconventional," provided they work [7].

Part I of the book you are currently reading asked how corporations can truly innovate to become both a force for good and remain a producer of profits in the future. Looking at the example of the textile and garment industry, Part II then examined how to upgrade global supply chains, the bleeding edge of the market economy in many countries. Part III now looks at the energy industry. Its progress is ultimately essential to enable higher competitiveness paired with better social and economic performance in the other industries that depend on reliable and affordable energy. As the future powers on, this century's key challenge is affordable, carbon-neutral (or carbon-negative) energy, as well as the security of its supply. Seen that breakthroughs are needed to achieve this goal, we need to be ambitious and ask: what is the next plow in the digital global age—the tool which made human civilization first possible? And what is the next—this time, zero emissions—steam engine?

Remember the 'solex' in the James Bond movie "The Man with the Golden Gun," released right after the first oil crisis? [8]. That's it. Daring is essential to locate inventions that can fast track our energy future. The following chapters draw on the insights coming from an ambitious nine-month global search for clean energy game changers I led via the Exergeia Project in 2014–2015 and follow-on work, which considered over 9000 startups and projects. In energy as well, it's time to take the next qualitative leap forward. If we want exponential progress, we need to be ready for disruption! The following chapters showcase how to achieve this [9].

8.2 Be Aware You Are Dealing with a Political Animal

Anyone who wants progress is well advised not to be naïve about the forces backing the status quo. For nations and regions, access to energy is a question of competitiveness and geostrategic positioning. Take the EU, the largest energy importer in the world: the European Union (EU) imports 53 % of its energy, spending around EUR 400 billion per year [10]. The European Commission talks about establishing energy as the EU's "fifth freedom"—next to the free movement of goods, people, services and capital, which are all central to the EU and its global competitiveness [11].

On the demand side, there is work to do as well: 75 % of the EU's housing stock is considered energy inefficient. 94 % of transport relies on oil products (90 % of which are imported) [12]. Important for businesses and consumers, wholesale electricity prices in Europe are 30 % higher and wholesale gas prices over 100 % higher than in the US [13]. To secure its long-term competitiveness, Europe needs to solve this—and this will cost money. Estimates indicate that over EUR 1 trillion need to be invested into the EU energy sector by 2020 alone [14].

To complicate the picture, the transition toward clean energy starts from a low baseline and demand for energy keeps rising. In 2010, world total primary energy supply (TPES) was 13,371 million tons of oil equivalent (MTOE) [14]. About 13 % was produced from renewable energy sources (excluding nuclear power). This share has remained unchanged since 2000, but with changing contributions of the different renewable sources. Global energy consumption is expected to rise by about 45 %, with a growth averaging 1.5 % per annum from 2012 to 2035 [15]. Fully 95 % of the growth in demand is expected to come from the emerging economies. Energy use in the advanced economies of North America, Europe and Asia as a group is expected to grow only very slowly—and begin to decline in the later years of the forecast period. In other words, the centers of growth are shifting, but energy demand keeps growing. As it has thus far, with a 52 % rise in demand over the last twenty years, and a 30 % rise over the last ten years [15].

Further accentuated by global warming—which to be addressed will require retiring fossil energy sources on a massive scale over the next two decades—it is fair to say that the world of energy is now again in the middle of a transition. This results in both creative destruction and fresh opportunities. For example, the until recently stable and profitable business models of major utilities such as German energy giants E.ON or RWE—which relied on fossil fuels and oligopolistic market structures—are becoming obsolete. Without some breakthrough, and taking into consideration the energy landscape and current trends, the pace of rise of clean energy will be too feeble to mitigate global warming, or even to provide access to the 1.3 billion people without access to energy and to the 2.6 billion without access to clean cooking facilities [16, 17]. We are moving onto the fast lane toward the overshoot and collapse scenario that the Club of Rome had drawn attention to, and urgently need real solutions. Given the inherently political and highly regulated nature of energy markets, it is as important to create the right rules and regulations as it is to come up with the enabling technologies that can deliver on our needs in terms of price and performance.

8.3 Generation, Storage, Transmission: Invent to Disrupt

Wherever we may stand ideologically, let's be clear: turning back the clock is not possible. The only logical way forward to solve the energy problem is to invent our way out of the bottleneck we are racing toward—just as humanity has always successfully addressed the obstacles standing in its development path, until now. What needs to happen is known; the estimated costs have also been mapped [18]. It is however less obvious how a full energy transition can happen fast. To succeed, we need to take our cues from science—and be ambitious.

While many people like to discuss high-flying clean energy goals, fewer are looking at the details that will enable us to reach any ambition level that may be set. The good news is that amazing breakthroughs are in the wings, waiting to move from the lab to becoming commercial products. Take silicene, also dubbed "graphene's cousin," and made of one-atom thick sheets of silicon atoms. In the mid-2000s, scientists theorized that silicon atoms could form sheets similar to graphene—or pure carbon in the form of a one atom thick, nearly transparent sheet that is about 100 times stronger than steel, and that in spite of its low weight, conducts heat and electricity with high efficiency [19]. However, even if graphene is the world's most conductive substance, it is missing a crucial characteristic: unlike semiconductors, which are used in computer chips, graphene misses a band gap. This is the energy hurdle that electrons must overleap before they can carry current, thereby enabling semiconductor devices to switch on and off, performing 'logic' operations on bits. Now given its properties, if silicene could be used to build electronic devices, it could enable the semiconductor industry to achieve the Holy Grail in miniaturization. Until three years ago, this could have been a theorist's fantasy. However, in 2012, Guy Le Lay, a French materials scientist at Aix-Marseille University, managed to create silicene in the lab.

Many developments in chemistry, nanotechnology, and material science allow ideas from theoretical physics to be translated into practice. We need to bring them to bear on solving the problem, and fast. If we are serious about the energy transition, we need to make progress along all energy frontiers: energy generation, storage, transmission and efficiency. In short: disruption is the name of the game. The transition to an 80–100 % renewable economy over the next 35 years can only succeed if solutions move into place for all areas in which progress is needed, namely:

- Energy use/waste reduction (energy efficiency);
- Energy generation (considering all possible sources of renewable energy, but excluding nuclear power which is neither safe nor unlimited nor clean when waste disposal is considered);
- Energy storage;
- Energy transmission; and
- Energy distribution to end consumers.

8.4 LENR et alia: Let a Thousand Flowers Bloom

Making the energy industry clean will require quite a number of different solutions to be put to work from different ends of the problem at the same time. Just thinking about incremental improvements is unlikely to be good enough. What could possibly be the limit when we think about cheap energy, available everywhere without adverse emissions? Can we conceive a world where there is no need for coal, and perhaps not even for power grids? Where homes, communities and offices could generate their own power through a device that has an energy density of up to 8,000,000 times that of today's combustion energy process?

For example, a proposed solution that has excited many scientists now for several generations are so-called "low-energy nuclear reactions" (LENR) [20]. This is a type of nuclear reaction theorized to occur at near room temperatures [21]. If it can be achieved it could play an important role in the energy transition by producing energy that is much cleaner, safer and cheaper than any other known source of energy. How would this work? LENR has been theorized as a phenomenon where unusual amounts of heat are released when certain metals such as nickel or palladium absorb hydrogen or deuterium in the presence of an external stimulus, such as heat or an electric current [22].

Given that LENR would be the Holy Grail of energy generation, it is not surprising that many scientific attempts are ongoing now to replicate and control the LENR effect. One player in seeking to revolutionize this technology is E-Cat, standing for "Energy Catalyzer." E-Cat is a device invented by Italian-American inventor Andrea Rossi: when combined, hydrogen gas, powdered nickel and undisclosed proprietary catalysts produce a large amount of heat through a LENR process inside a specially designed chamber [23]. According to Rossi, the cost of energy produced by the E-Cats will be at least one sixth of the cost of conventional sources. Every six months the E-Cat units would need to have their reaction chambers containing processed nickel and the catalyst replaced by licensed agents. The potential of this device was independently assessed by three European Universities to have about 1000 times more peak power than gasoline [24].

To get a sense of the order of magnitude, if this technology were to be applied in the automobile industry, with the cost equivalent of one tank of gasoline one could drive several thousand kilometers, without generating any kind of emissions. In 2014, US firm Industrial Heat acquired the rights to Andrea Rossi's E-Cat

Technology for USD 11 million [25]. Skeptical scientists have long ridiculed LENR however. Locating it in the category of pseudo-science, they argued that laws of physics as we currently understand them do not permit a LENR device to work.

The jury is not yet in though. Let us not forget that immediately before the advent of quantum physics one hundred years ago, leading scientists argued: "All that is left in physics is to figure out the sixth zero behind the decimal" [26]. Once quantum physics burst onto the scene, we learned that this assessment was false. Dennis Bushnell, chief scientist at National Aeronautics and Space Administration's (NASA) Langley Research Center argues that LENR has "demonstrated ability to produce excess amounts of energy, cleanly, without hazardous ionizing radiation, without producing nasty waste" [27]. According to Bushnell, estimates suggest that 1 % of the nickel mined on the planet each year could produce the world's energy requirements at 25 % of the cost of coal [27].

Currently, there are however no viable LENR devices on the commercial market. The bulk of research money is going elsewhere. For example, projects such as "International Thermonuclear Experimental Reactor" (ITER, known as the "the way" in Latin) indicate a continued general interest in nuclear fusion: the EU-led international 500 MW nuclear fusion research project is currently building the world's largest experimental tokamak nuclear fusion reactor. Located next to the Cadarache nuclear research facility in Southern France, the USD 16 billion project —financed by the seven parties to ITER, the EU, the US, Russia, Japan, China, South Korea and India—seeks to use plasma physics to achieve a full-scale electricity-producing fusion power plant [28].

8.5 It's the Infrastructure, Stupid

Next to the ability of a technology to provide enough cheap renewable energy in a narrow sense, the structure of the energy landscape in an economic region sets the framework conditions on what's possible. In spite of Europe's historically leading role in investment in renewable energy, European renewable energy businesses are still small in comparison with fossil fuel companies: in 2014, they had a combined annual turnover of EUR 129 billion and employed just over a million people [29, 30]. EU greenhouse gas emissions fell 19.2 % from 1990–2012, benefiting from the demise of polluting industries following from the transformation of Eastern Europe after the dismantlement of the Soviet bloc. Partially driven by policy goals, the renewables industry is set to grow: by 2030, the EU aims to cut greenhouse gas emissions by at least 40 %, boost renewable energy by at least 27 %, and improve energy efficiency by at least 27 % [30]. This is also the EU's vision for a global climate agreement to be negotiated in Paris in December 2015.

The new regulatory idea on the table is the goal of an energy union, or termed more officially, a "Framework Strategy for a Resilient Energy Union with a Forward-Looking Climate Change Policy" [31]. To achieve the Energy Union, the

EU Commission plans to introduce new legislation to redesign the electricity market. Other goals include ensuring more transparency in gas contracts, and developing regional cooperation as an important step towards an integrated market. In practice, this means a stronger regulatory framework, new legislation to ensure the supply for electricity and gas, and increased EU funding for energy efficiency or renewables. Provided public funds are used to finance and provide market access for the most promising inventions, the Energy Union could help to significantly fast track the energy transition. Let's hope that regulation will not be such that it drives out private initiative. A common criticism of accelerating the energy transition is that the grid will not be able to handle a very large volume of intermittent sources of power, so grid stabilization is necessary. For example, in discussions one hears routinely from its neighbors that Germany destabilizes the grid—because it is pressing ahead with an ambitious energy transition. Tackling the technical constraint that will otherwise impose a ceiling on the growth rates of renewables will precisely require fresh solutions.

Even otherwise well-integrated economic areas are not necessarily fully integrated grid-wise. For example, in spite of its ambitious plans, the EU is not even close to a single homogeneous energy market. Defined as at least 10 % of installed electricity production capacity being able to "cross borders," 12 EU member states do not yet meet the EU's minimum interconnection target (namely Cyprus, Estonia, Ireland, Italy, Latvia, Lithuania, Malta, Poland, Portugal, Romania, Spain and the UK) [32]. Ten percent electricity interconnection by 2020 is the minimum necessary for electricity to flow and be traded among EU states [32]. An appropriately interconnected European energy grid could save consumers up to EUR 40 billion a year [32]. If the EU implements its 137 listed electricity projects—including 35 on electricity interconnection—only two member states would remain under-connected. To take a significant step forward, we need better solutions to both energy storage and to electric power transmission, or the bulk transfer of electrical energy from the point of generation to electrical substations that are located near demand centers. A related question is what the appropriate mix between centralized and decentralized power generation would be in a 100 % renewable economy. Let us have a look at what more established technologies than LENR have to offer in this respect.

References

1. Meadows DH et al (1972) The limits to growth. Universe Books, New York
2. Mizroch A (2014) Charge your phone in 30 seconds? An Israeli firm says it can. http://blogs.wsj.com/digits/2014/04/07/charge-your-phone-in-30-seconds-an-israeli-firm-says-it-can. Accessed 17 Mar 2015
3. Clark P, Crooks E (2015) European energy groups seek UN backing for carbon pricing system. http://www.cnbc.com/id/102720664. Accessed 1 Jun 2015
4. Butler N (2015) How oil and gas majors are rethinking on climate change. http://blogs.ft.com/nick-butler/2015/04/06/oil-and-gas-majors-are-rethinking-on-climate-change. Accessed 6 Apr 2015

5. Quaschning V (1999) Renewable energy and climate change. http://www.scribd.com/doc/38412962/Renewable-Energy#scribd. Accessed 22 Mar 2015
6. Meadows DH et al (1972) The limits to growth. Universe Books, New York
7. Wikiquote (2015) Deng Xiaoping. http://en.wikiquote.org/wiki/Deng_Xiaoping. Accessed 2 Apr 2015
8. In the James Bond movie "The Man with the Golden Gun," the solex agitator is a small component made by a British scientist for an ultra-high efficiency solar device as a solution to the 1973 energy crisis. Geocoaching (2014) Bond objects: solex agitator. http://www.geocaching.com/geocache/GC52JZR_bond-objects-solex-agitator. Accessed 15 Jul 2015
9. Martin M (2010) After Copenhagen: perspectives on energy. http://ssrn.com/abstract=1532825. Accessed 9 Aug 2014
10. Greenpeace (2014) The EU energy (r)evolution. http://www.tagesschau.de/greenpace-energie-100.pdf. Accessed 22 Mar 2015
11. IEA (2011) National electricity boards. http://shrinkthatfootprint.com/average-electricity-prices-kwh. Accessed 22 Mar 2015
12. European Commission (2015) Energy Union: secure, sustainable, competitive, affordable energy for every European. In: European Commission—Press release http://europa.eu/rapid/press-release_IP-15-4497_en.htm. Accessed 22 Mar 2015
13. Eteris E (2015) New EU energy and climate strategy: fifth basic freedom. In: The Baltic course. http://www.baltic-course.com/eng/energy/?doc=102980. Accessed 22 Mar 2015
14. IEA (2014) The IEA register. http://www.iea.org/publications/freepublications. Accessed 22 Mar 2015
15. BP (2015) BP Energy Outlook 2035. http://www.bp.com/en/global/corporate/press/press-releases/energy-outlook-2035.html. Accessed 22 Mar 2015
16. United Nations Foundation (2013) Achieving Universal Energy Access. http://www.unfoundation.org/what-we-do/issues/energy-and-climate/clean-energy-development.html. Accessed 22 Mar 2015
17. IEA (2015) Energy Poverty. http://www.iea.org/topics/energypoverty. Accessed 8 Apr 2015
18. Substantive work on this topic has been conducted by the strategy consultancy McKinsey&Company, including the estimate of a widely referenced cost curve for greenhouse gas reduction. Nauclér T et al (2007) A cost curve for greenhouse gas reduction. http://www.mckinsey.com/insights/sustainability/a_cost_curve_for_greenhouse_gas_reduction. Accessed 22 Mar 2015
19. Martin M (2015) Energy transition fast forward: the exergeia manifesto. http://impacteconomy.com/en/articles2015.php. Accessed 17 Mar 2015
20. LENR Proof (2013) Is LENR the real deal? http://www.lenrproof.com/slide_02.html. Accessed 3 Mar 2015
21. Marvis D, Wells D (2015) The application of LENR to synergistic mission capabilities. NASA & Georgia Institute of Technology. http://ntrs.nasa.gov/archive/nasa/casi.ntrs.nasa.gov/20150000549.pdf. Accessed 5 Mar 2015
22. LENR Proof (2013) Is LENR the real deal? http://www.lenrproof.com/slide_02.html. Accessed 3 Mar 2015
23. E-Cat World (n.d.) What is the E-Cat? http://www.e-catworld.com/what-is-the-e-cat. Accessed 4 Mar 2015
24. Levi G et al (2013) Indication of anomalous heat energy production in a reactor device containing hydrogen loaded nickel powder. http://arxiv.org/ftp/arxiv/papers/1305/1305.3913.pdf. Accessed 11 Mar 2015
25. PRNewswire (2014) Industrial heat has acquired Andrea Rossi's E-Cat technology. http://www.prnewswire.com/news-releases/industrial-heat-has-acquired-andrea-rossis-e-cat-technology-241853361.html. Accessed 8 Apr 2015
26. Michelson A (1894) Speech at the dedication of Ryerson Physics Lab. University of Chicago. http://amasci.com/weird/end.html. Accessed 4 Mar 2015

27. McMahon J (2013) NASA: a nuclear reactor to replace your water heater. In: Forbes. http://www.forbes.com/sites/jeffmcmahon/2013/02/22/nasa-a-nuclear-reactor-to-replace-your-water-heater. Accessed 7 Mar 2015
28. ITER (2015) The ITER register. http://www.iter.org. Accessed 18 Mar 2015
29. Eteris E (2015) New EU energy and climate strategy: fifth basic freedom. In: The baltic course. http://www.baltic-course.com/eng/energy/?doc=102980. Accessed 8 Mar 2015
30. Ren21 (2014) Renewables 2014 global status report. http://www10.iadb.org/intal/intalcdi/PE/2014/14403.pdf. Accessed 8 Mar 2015
31. European Commission (2015) A framework strategy for a resilient Energy Union with a forward-looking climate change policy. http://ec.europa.eu/priorities/energy-union/docs/energyunion_en.pdf. Accessed 8 Mar 2015
32. European Commission (2015) Energy Union: secure, sustainable, competitive, affordable energy for every European. In: European Commission—Press release http://europa.eu/rapid/press-release_IP-15-4497_en.htm. Accessed 8 Mar 2015

9 Technologies for the No Subsidy Imperative

9.1 Solar PV: Efficiency Revolution, Now

If the energy transition is to succeed, subsidies can be deployed only to reach critical mass, rather than to operate at scale. Total competitiveness is a must. To compete, renewable energy will have to become fully price competitive and substitutable with fossil fuels much faster. If renewables can turn out electricity at a few cents a kWh and storage solutions can compete with gasoline in terms of energy density and ability to release energy, then we are talking. In principle, if we leverage science, solar panels could even capture energy at night (though that is more like a gimmick and alone will not solve the energy crisis).

The only way to achieve this is through science and innovation. Can this be done? Well, in the past, scientists have achieved equivalent advances. The next generation of solutions brings current advances in fields such as materials sciences, nanotechnology, information technology, or engineering to bear on the problem.

The efficiency improvements in solar photovoltaics over the past 40 years and current progress point to an efficiency revolution in the wings.[1] Multi-junction cells allow for a 46 % conversion efficiency today [2]. And the fast rise of the efficiency curve of 'perovskites,' a newer class of materials with a particular crystal structure is exciting: their band gap can be tuned by changing halide content to become very good light absorbers over the whole visible solar spectrum. Reported high carrier mobility and long diffusion lengths mean that photo generated charges can travel longer distances, increasing charge absorption and ultimately producing more electricity. Moreover, using rectifying antennas (or 'rectennas'), one can already convert electromagnetic radiation to electricity, with reported efficiencies of over 90 % in the microwave range.

[1]For a chart summarizing solar research cell efficiency improvements from 1976–2015, see [1].

M. Martin, *Building the Impact Economy*,
DOI 10.1007/978-3-319-25604-7_9

In solar photovoltaics, there are two especially promising directions:

First, physics predicts that rectennas can also reach microwave range efficiencies in the infrared and optical ranges, converting sunlight to electricity. An exciting step forward follows then from the energy cost and availability implications of high-efficiency solar energy harvesting, which leverages the full solar radiation spectrum during the day. Possibly one can even conduct long wave radiation energy harvesting at night.

A second breakthrough opportunity relates to trapping sunlight in solar cells much better. Optical losses consist of light that could have generated an electron-hole pair, but does not, because the light is reflected from the front surface or because it is not absorbed in the solar cell [3]. Optical losses chiefly affect the power from a solar cell by lowering the short circuit current. The way forward is an optical (nano) antenna that keeps the light trapped longer in the cell than by previously used microstructures. For example, Nanotechnology Solar, a startup based in Germany has come up with a solution of using advanced nano-imprint lithography that allows the cost-effective production of gradient nanostructures for optical components. First preliminary experiments and simulation results indicate that this solution could increase the relative efficiency of solar cells by over 10 %; in the future, a silicon solar cell/module with a relative efficiency of over 24 % could be realized [4].

9.2 Get "Big Wind 2.0" Blowing

The Earth's surface consists of different types of land and water bodies and absorbs the sun's heat at different rates. The uneven heating causes the movement of air from an area of high pressure to an area of low pressure. Created by this uneven heating of the Earth's surface by the sun, wind energy is one of the cleanest, most environmentally friendly energy sources. Wind turbines produce mechanical or electrical power from airflow. The potential of wind power is enormous, amounting to nearly 20 times more than what the entire human population needs [5, 6].

The rise of the modern wind industry using small wind and large onshore and offshore turbines producing electricity by rotating around a horizontal or vertical axis dates back to the 1980s. Windmills—their predecessors—have been around for millennia and wind generators for electricity have been used on farms since the 1930s [7]. The world's biggest wind turbine in 2015, the Vestas V164, was a 8 MW geared offshore turbine that came on stream in January 2014, with 80 meter-long blades and a lightweight nacelle; it is currently being tested at the Danish national wind turbine test center at Osterild [8]. Large 1 MW wind turbines can generate enough electricity to supply about 600 US homes. Industry experts predict that wind will supply one third of the world's electricity needs by 2050 [9].

Today, there are several interesting frontiers in wind power that could help to further harness its potential (Table 9.1), relating to:

- Horizontal axis wind turbines;
- Vertical axis wind turbines;
- Bladeless turbines;

Table 9.1 Performance and readiness of wind technologies

Technology	Description	Breakthrough	Environmental impact	Challenges	Technological readiness
Horizontal axis turbines	The most widespread technology used both on- and offshore	Breakthroughs have led to larger systems and offshore capabilities	High materials footprint; Visual environmental impact	As a mature technology, the challenge is cost reduction	Mature
Vertical axis turbines	Wind generators with blades spinning around a vertical axis	Smaller swept area resulting in reduced wake effect can receive wind from any direction	High materials footprint; Visual environmental impact	Lower efficiency than horizontal axis turbines	Mature
Bladeless	Bladeless tower can generate energy by vibrations caused by wind	By eliminating the need for blades, materials and maintenance cost are greatly reduced	Less materials translates into a 40 % lower CO_2 footprint; Bird kills reduced as well	The system would need to be much taller to produce the same energy as a conventional system	R&D/early prototype
High-altitude wind (HAW)	Wind's kinetic energy increases as cube function of wind speed which greatly increases with altitude	Due to breakthroughs in lightweight materials and sensors, high altitude kites can fly autonomously	By eliminating concrete and steel structures, HAW can use 90 % less material	Autonomous takeoff and landing, regulatory challenges	Several companies have working prototypes
Bionic/nano coating	Through biomimicry, researchers are applying designs perfected by nature over millions of years	By imitating feathers for example, turbines could operate in highly variable winds and with greater efficiency	By increasing efficiency, wind resources are maximized	Cost effective production and durability	R&D

Source Author [10]

- High-altitude winds; and
- Bionic coating as well as coating using nanotechnologies.

In assessing these opportunities, it is helpful to ask how they promise progress in the wind industry associated with one or more of four themes:

- Dealing with the wake effect;
- Repowering;
- Addressing atmospheric turbulence, and
- Harnessing high-altitude winds (HAW).

First, the so-called "wake effect" is an aerodynamic turbulence effect that occurs between commercial upwind turbines and downwind turbines, changing natural wind speeds [11]. Upwind turbines create wind wakes that affect the natural wind flow to adjacent downwind turbines. This then causes the downwind turbines to experience diminished energy production and in some cases, increased mechanical loads, leading to overall power losses.

Any wind turbine design that can significantly reduce exposure to the wake effect and run at lower wind speed is relevant to boost economics, power production and land usage efficiency. For example, Swiss company Bogga Wind Power has an advanced design for a vertical axis wind turbine that may be able to overcome the wake effect present in today's large wind farms. As a result, their wind turbines are expected to have a higher power generation efficiency, produce lower noise (0–10 dB compared to 5–60 dB from a horizontal axis). They also increase land efficiency and are able to harvest turbulent and higher wind speeds [12].

Second, "repowering" or the process of replacing old turbines with modern wind turbines is becoming a necessity because of both age and performance. For investors in Germany, repowering got a boost with the amendment of the Renewable Energy Law (EEG) on January 1, 2009. Operators of older wind farms came to profit not only from the higher yield of new wind turbines, but also from the so-called repowering bonus (at the rate of 0.5 EUR cent/kWh), an additional modernization incentive [13]. Unfortunately, the German regulators dismissed the repowering bonus introduced in 2009. Moreover, in 2017, Germany will stop the feed-in-tariff, and introduce project tendering. This will put additional pressure on wind turbine manufacturers to reduce overall wind power project cost. The new regulation will impact the financing of new wind farms. These are typically located on low wind speed sites, because high wind speed sites are already developed. This brings additional importance to the repowering of aging wind turbines that are already installed on the best wind resources and accentuates the need for maximizing land usage efficiency (i.e., reducing the wake effect).

Third, technologies that can harvest winds at higher altitudes are interesting: these winds become steadier, more persistent, and have a higher velocity and thus more kinetic energy. For example, for harvesting high-altitude winds, Italian company Kitenergy has come up with a groundbreaking concept of using airfoils or power kites, similar to those used for kite surfing. Linked to the ground by two cables, the power kite reaches altitudes of about 600–1000 m above the ground. Winds are stronger at such elevations and more constant compared to the elevation where current wind towers operate. The flight of the airfoil is tracked using on-board wireless instrumentation and driven by an automatic control unit, able to differentially pull the lines to influence the wing motion. By converting the traction forces acting on the airfoil, electricity is generated at ground level by rotating mechanisms and electric generators. Electric drives, drums, the on-board sensors and all the hardware needed to control a single kite form a system, the "Kite Steering Unit" (KSU), which forms the core of the Kitenergy technology.

A benefit of electricity generation at the ground level is the reduction of weight: a typical 2 MW conventional wind turbine weighs 300 tons, whereas an otherwise equivalent Kitenergy system would weigh only two tons. The company forecasts that they will be able to set up a 100 MW wind farm at a levelized cost of energy (LCOE) significantly less than EUR 50/MWh [14]. This compares favorably to the US Energy Information Administration's (EIA) forecast for the lowest LCOE in 2019 of a fossil fuel electric generation of EUR 51.7/MWh [15].

Another constraint standing in the way of Big Wind 2.0 is atmospheric turbulence. Atmospheric turbulence impacts wind energy through power performance effects with adverse impacts on turbine loads and noise propagation. To respond, Munich based startup AF-AX has come up with an innovative adaptive rotor blade technique based on bionics that can improve aerodynamics of the blade and help overcome atmospheric turbulence [16]. Coating wind turbines so they can deal much better with turbulence and produce energy at lower wind speeds, as well as suffering from much less downtime is in principle possible—merging design thinking from bionics with advances in materials sciences.

9.3 Linear Progress Ain't Good Enough: What's Up with Storage?

We can see that relevant progress could be around the corner in solar and wind. Notwithstanding, renewable sources produce energy only intermittently. As a result, storage becomes a key constraint for the energy transition; the sun does not always shine and the wind does not always blow; but the demand for energy remains. Energy storage thus plays a key role to integrate more renewables into the grid. We need to achieve continuous and flexible supply of energy and reduce energy costs by peak shaving [17]. A breakthrough in batteries is moreover a prime enabler of e-mobility. Alas progress in storage compares very unfavorably with, say, advances in semiconducting: rather than progress on a logarithmic scale in computing power, we are looking at progress on a linear scale in storage. That needs to change: the need of the hour is a technology that is capable of effectively maximizing the output, reducing the size and weight of current systems, increasing performance and reducing life cycle costs for batteries—all while improving safety and stability under harsh conditions. The 2022 goals set by the US Department of Energy with regard to batteries and energy storage provide a useful guideline for identifying such technologies [18]:

- Reduce the production cost of an electric vehicle (EV) battery to a quarter of its current cost;
- Halve the size of an electric vehicle battery; and
- Halve the weight of an electric vehicle battery;

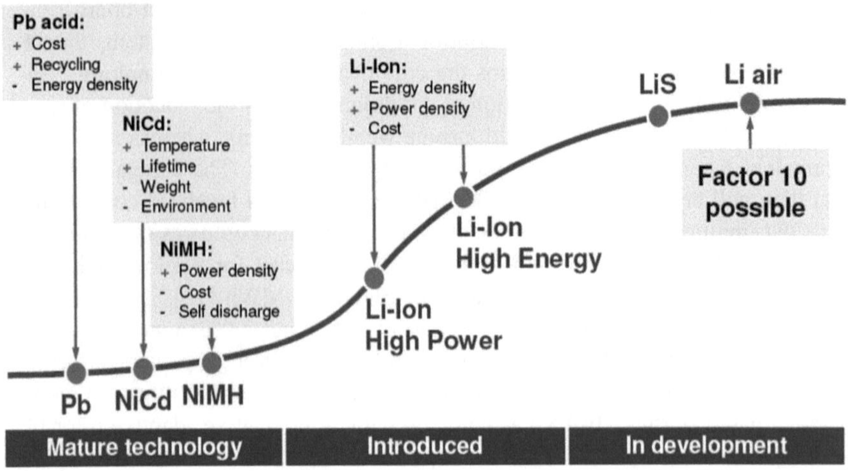

Fig. 9.1 Battery technology roadmap (*source* Gopalakrishnan et al. [19]) (Figure courtesy of Huib van Essen [19])

- Achieving these goals would result in lowering battery cost from USD 500/kwh to USD 125/kwh; and
- Increasing density from 100 to 250 Wh/kg.

Currently, electric vehicles use a large quantity of Li-ion batteries. These batteries suffer from many drawbacks such as cost, overweight, limited capacity, reliability and risk of fire (remember when an empty Japan Airlines 787 Dreamliner caught fire in January 2013 while parked at Boston's Logan Airport because of a battery cell short-circuit?). This makes them less attractive than standard fossil fuel vehicles. New combinations such as Lithium and Sulfur hold the potential to dramatically raise batteries' energy density and lower their cost (Fig. 9.1).

For example, UK-based OXIS Energy Ltd. has developed a unique Sulfur based cathode material with high stability and having a theoretical specific energy in excess of 2700 Wh/kg, which is nearly five times higher than that of Lithium-Ion Batteries. OXIS expects to come out with a prototype in 2016 that will have an energy density of 400 Wh/kg with over 2000 cycles before the capacity reduces to 80 % of the energy density at its beginning of life. The battery is projected to cost less than USD 250/kWh [20]. To improve storage, pursuing several pathways is essential, but with a shared goal: roll out a technology whose energy density can compete with gasoline.

References

1. NREL (2015) Best research-cell efficiencies. http://www.nrel.gov/ncpv/images/efficiency_chart.jpg. Accessed 10 Aug 2015
2. Colthorpe A (2014) Soitec-Fraunhofer ISE multi-junction CPV cell hits world record 46 % conversion efficiency. http://www.pv-tech.org/news/soitec_fraunhofer_ise_multi_junction_cpv_cell_hits_world_record_46_conversi. Accessed 5 Mar 2015

3. Photovoltaic Education Network (2015) Optical losses. http://www.pveducation.org/pvcdrom/design/optical-losses. Accessed 5 Mar 2015
4. Nanotechnology Solar (2015) Project teaser NTS, Germany
5. Kimberly D, Crivella E (2011) Wind turbine wakes, wake effect impacts, and wind leases: using solar access laws as the model for capitalizing on wind rights during the evolution of wind policy standards. http://scholarship.law.duke.edu/cgi/viewcontent.cgi?article=1192&context=delpf. Accessed 5 Mar 2015
6. Choi C (2012) New research demonstrates that wind power could generate all the world's electricity needs without large atmospheric effects. http://www.insidescience.org/content/studies-show-wind-powers-massive-potential/782. Accessed 5 Mar 2015
7. Union of Concerned Scientists (2015) Farming the wind: wind power and agriculture. http://www.ucsusa.org/clean_energy/smart-energy-solutions/increase-renewables/farming-the-wind-wind-power.html#.VQmnOo7F8YF. Accessed 5 Mar 2015
8. De Vries E (2013) Denmark: close up-Vestas V164-8.0 nacelle and hub. http://www.windpowermonthly.com/article/1211056/close—vestas-v164-80-nacelle-hub. Accessed 5 Mar 2015
9. National Geographic (2015) Wind power. http://environment.nationalgeographic.com/environment/global-warming/wind-power-profile/. Accessed 5 Mar 2015
10. Martin M (2015) Energy transition fast forward! Scouting the solutions for the 80–100 % renewable economy, p 18. Impact Economy. http://www.impacteconomy.com/papers/IE_PRIMER_APRIL2015_EN.pdf
11. Kimberly D, Rivella E (2011) Wind turbine wakes, wake effect impacts, and wind leases: using solar access laws as the model for capitalizing on wind rights during the evolution of wind policy standards. http://scholarship.law.duke.edu/cgi/viewcontent.cgi?article=1192&context=delpf. Accessed 5 Mar 2015
12. BoggaWind (2015) Presentation BoggaWind, Switzerland
13. IEA (2012) Combined heat and power law Germany. http://www.iea.org/policiesandmeasures/pams/germany/name-22083-en.php. Accessed 2 Mar 2015
14. NREL (2013) Levelized cost of energy (LCOE). https://www.nrel.gov/analysis/sam/help/html-php/index.html?mtf_lcoe.htm. Accessed 2 Mar 2015
15. IEA (2014) Renewable energy medium-term market report. http://www.iea.org/Textbase/npsum/MTrenew2014sum.pdf. Accessed 5 Mar 2015
16. AF-AX (2015) The AF-AX register. http://www.af-ax.com. Accessed 22 May 2015
17. Energy.gov (2015) Vehicle technologies office: batteries. http://energy.gov/eere/vehicles/vehicle-technologies-office-batteries. Accessed 8 Mar 2015
18. Energy.gov (2014) EV everywhere grand challenge. http://energy.gov/sites/prod/files/2014/02/f8/eveverywhere_road_to_success.pdf. Accessed 5 Mar 2015
19. Gopalakrishnan D, van Essen H et al (2011) Assessment of electric vehicle and battery technology, p 12. http://ec.europa.eu/clima/policies/transport/vehicles/docs/d2_en.pdf. Accessed 10 Nov 2014
20. Oxis Energy (2014) Next generation battery technology: http://www.oxisenergy.com. Accessed 5 Sept 2014

Rebreed Field Mice into 800-Pound Gorillas

10

10.1 Renewables: Take off from an Underwhelming Baseline

If we assume we have the relevant technologies, the next question is how to get them deployed at scale. One thing is for sure: "Scale-up-as-usual" is not good enough: the current growth projections for renewable energy in the overall global energy mix are simply too unambitious. For example, the BP Global Energy Outlook 2015 predicts that among non-fossil fuels, renewables including biofuels but excluding hydro and nuclear power will advance from a market share of 3 % today to around 8 % in 2035, overtaking nuclear energy in the 2020s and hydropower in the 2030s. In electricity generation, the percentage of renewables is much higher already: BP predicts that growing renewable energy supply will lift the aggregate non-fossil share from 32 % in 2013 to 38 % in power generation by 2035 [1]. Due to idiosyncratic factors, some countries already have a high share of renewables in power generation. For example, in New Zealand 70 % of all electricity comes from renewable sources already, mainly from hydropower and geothermal power [2].

The crucial *how-to* questions follow readily: (1) Which technologies do we need to bring to market to raise the share of renewables to 80+ % in the energy mix, and close to 100 % in power generation? (2) Which critical blockers stand in the way of a large-scale transition to intermittent sources of renewable energy? To get a sense of the size difference between David and Goliath, let us first have a look at the baseline.

Compared to their inherent potential, the penetration of renewables in today's energy mix is underwhelming. Let's have a look at where the different technologies stand today.

Renewable energy is energy that is derived from natural processes (e.g. sunlight and wind) that are replenished at a higher rate than they are consumed. Solar, wind, geothermal, hydropower, bioenergy and ocean power are all sources of renewable energy. The share of traditional biomass out of total renewable energy fell from 50 % in 2000 to 45 % in 2010, while biofuels (transport fuels produced from

M. Martin, *Building the Impact Economy*,
DOI 10.1007/978-3-319-25604-7_10

biomass feedstock) met a growing share of transportation fuel needs [3]. Hydropower, the largest source of renewables-based electricity, remained stable.

Hydro is currently the largest renewable source for power generation in the world, producing 3431 Terawatt-hours (TWh), meeting 16.3 % of global electricity needs at competitive prices and representing about 85 % of global renewable electricity [4]. Hydro dominates the electricity mix in several countries (industrialized, emerging or developing); 159 countries have a fully mature hydro technology. Global hydropower capacity is projected to increase from 1067 Gigawatt (GW) in 2011 to over 1680 GW in 2035; incremental hydropower production in OECD countries, where the best resources have already been exploited, is however limited [5].

On the electricity front, there is progress. Electricity does not fulfill all our energy needs though. Electricity generation from wind grew by 27 % and solar photovoltaics by 42 % per year on average from 2000 to 2010 [5]. The role of renewable sources in the global power mix, in particular, continues to increase rapidly. On a percentage basis, renewables continue to be the fastest-growing power source. As global renewable electricity generation expands in absolute terms, it is expected to surpass that from natural gas and double that from nuclear power by 2016, becoming the second most important global electricity source, after coal [6]. On the current transition path, global renewable generation is estimated to rise to 25 % of gross power generation in 2018, up from 20 % in 2011 as deployment spreads out globally [6]. Absent a few exceptions, we are nevertheless not nearing 80–100 % power generation by renewables. And even "renewables as usual" is costly. Under the current gradualist approach, estimated investments in renewables of USD 6.4 trillion are required over the period 2012–2035 [7]. The power sector is estimated to account for 94 % of the total, including wind (USD 2.1 trillion), hydro (USD 1.5 trillion) and solar photovoltaics (USD 1.3 trillion), with the remainder in biofuels [7].

Another theater is biomass. Due to its widespread non-commercial use in developing countries, bioenergy is the single largest renewable energy source today, providing 10 % of the world's primary energy supply [7]. It plays a crucial role in many developing countries, where it provides basic energy for cooking and space heating, but often at the price of severe health and environmental impacts. The deployment of advanced biomass cook stoves, clean fuels and additional off-grid biomass electricity supply in developing countries are key measures to improve the current situation and achieve universal access to clean energy facilities by 2030. Global primary energy demand for bioenergy, excluding traditional biomass, is expected to more than double from 526 MTOE in 2010 to nearly 1200 MTOE by 2035, growing at an average rate of 3.3 % per year [7].

When people hear renewables, they often think wind or solar. Solar photovoltaics, solar thermal electricity and solar heating and cooling are well-established solar technologies. The data is nevertheless sobering. Today, solar energy produces only a small fraction of the TPES. On the projected path, it is expected to rise to just over 2 % of total electricity generation in 2035. The potential is much greater though if better solar technologies can come on stream, thus enabling more investment.

Solar photovoltaics, often considered the poster child of the renewables industry, decreased from a record USD 91.6 billion in 2011 to USD 79.7 billion in 2012, as continued growth in annual capacity additions was not enough to offset falling photovoltaics prices, but is forecast to reach USD 123.7 billion towards 2022 [7]. Raising this investment path by several orders of magnitude in ways that make economic sense is a must if we want to seriously accelerate the energy transition.

Wind offers a similar story. Wind energy is considered to develop towards a mainstream, competitive and reliable power technology; yet it currently only produces 2.5 % of global electricity demand [8]. Globally, progress continues to be significant, with countries that are more active and players increasing annual installed capacity and investments. Technology improvements have continuously reduced energy costs, especially on land; the industry has overcome supply bottlenecks and expanded supply chains. Wind is projected to grow from USD 73.8 billion in 2012, up from USD 71.5 billion the previous year, to USD 124.7 billion in 2022 [8]. For both solar and wind technology, the challenge is: how can they graduate from their current status of field mice in the global energy mix and unlock the "800-pound gorilla" potential that is inherent in the sun and wind as sources of energy?

In biofuels there is work to do as well. Fuels are important, because today we cannot run everything on electricity. Investments in biofuels reached USD 95.2 billion in 2012, up from USD 83 billion the previous year, and are projected to grow to USD 177.7 billion by 2022 [8]. Again, this is understandable because we need the prospect of much better technologies to absorb more capital, which is currently too little to make a difference.

Other technologies in the mix are geothermal and ocean power. While their potential is too small to make a major global contribution to truly taking off from today's baseline, they can be important in terms of local or regional supply. Geothermal power comprises mature renewable technology options that can provide base-load power from energy stored in trapped vapor and liquids. In 2012, global geothermal power capacity was 11.4 GW and generated around 72 TWh of electricity [9]. Enhanced geothermal technologies are under development that would allow expanding the use of this technology family beyond countries that have resources suitable for established technologies. As a pioneer Iceland has a 25 % share of its total electricity demand provided by geothermal power [10]. Ocean power, with a minor capacity of 0.54 GW in 2012, encompasses five different types of technologies that exploit the following phenomena, none of which is widely deployed yet: tidal rise and fall (barrages), tidal/ocean currents (kinetic energy), waves, temperature gradients, and salinity gradients [11].

10.2 Build Innovations' Pathway to Market

There are many potential renewable technologies to create electricity, heat, or even fuel. However, even brilliant ideas rarely succeed on their own. They need pathways to dissemination. The same holds true for new energy technologies.

Significant barriers affect all innovation. But they are particularly challenging for the renewable energy sector. This results in a Research & Development (R&D) intensity as a percentage of value added that is close to zero [12]. For many firms, the proverbial wide "valley of death" is real: innovative technologies succeed in mobilizing funding to take them from observation of principles to experimental proof of concept—to only then being unable to receive adequate investment to push them through from development to commercialization.

So how can any promising technology overcome the primary barriers to adoption of clean energy technologies [13, 14]? This is where a sort of yin and yang becomes relevant: connecting a government's top-down work, a scientist's ability to invent, and an entrepreneur's ability to bring innovative products and services to market. Good strategies are needed to deal with the following four questions:

First, existing infrastructure encourages evolutionary rather than revolutionary technological innovation. How do we keep radically better new technologies from being blocked by prohibitively high capital costs required for new infrastructure?

Second, high upfront costs combined with limited access to finance deter investors by creating uncertainty about success prospects. They also raise capital cost and lengthen timescales of development. What is a project's plan to be as asset and capital light as reasonably possible?

Third, product differentiation can be difficult or impossible in the energy sector: for example, customers want and need electricity, but seen that electricity is a commodity, most may not care to find out whether it came from a clean and/or renewable source or not. So how can a clean energy technology compete on a price and performance only basis?

Fourth, how does one deal with spillover effects, which may benefit other firms and a society as a whole a lot more than they benefit the innovator (who needs to succeed on the open market where fast followers and substitutes are present)?

10.3 Let Governments Enable Breakthrough

Drawing on the report "Restart, Italia!" elaborated by a ministerial task force, new legislation in Italy recognized innovative startups in 2012 as an important driver to achieve sustainable growth, technological development and youth employment, as well as increasing social mobility, and attracting investments and talent from abroad. The plan took bold action by providing an assortment of strategic incentives to help these companies succeed and grow. Benefits for these innovative startups fall into the following six buckets [13, 14]: First, support in the form of cost saving exemptions related to company registration and corporate legal fees. Second, increased employment flexibility and the ability to hire external service providers, such as lawyers and accountants, through "work for equity" provisions. Third, tax incentives for companies and individuals for the years 2013–2016. These incentives apply both to direct investments and to other companies who invest primarily in startups. Incentives are even greater if the startup involves social goals or focuses on the energy sector. Fourth, introduction of crowdfunding to open capital sources from non-professional investors. Fifth, easy access to a Central Guarantee Fund that facilitates access to credit by offering guarantees on bank loans (incubators are also offered these privileges). Finally, support in international expansion including assistance with regulations, corporate tax, real estate, contracts, payment of expenses related to important international trade fairs and events, and most importantly, matching the startups with potential investors.

The venture capitalists who provide capital for commercialization of startups such as the ones incentivized by the Italian Government are not in the business of formulating public policy. Notwithstanding, innovation activity can be incentivized or dis-incentivized by public policy. Next to market size and maturity, as well as general ease of doing business, this adds an important framework condition.

Governments can fast track renewable energy market development by providing intelligent policies to push technological development with public funding. Moreover, they can accelerate technology adoption in the market by creating conditions that seek to reflect their value added (including environmental externalities), and by removing barriers to entry.

For an investor seeking to bring a groundbreaking product to market, the openness of the market matters. For example, in recent years the UK has worked on creating an environment where utilities are adopting low-carbon technologies, and turning them into useful products. This demand driven approach contrasts with continental Europe's general primacy of the supply side, funding research that leads to products without opening up the pathway to the market. As a result, although continental Europe is well positioned in the clean energy sector, both for R&D and industry, it has not been able to capitalize on its investments as well as it could—whereas others are now piggybacking on its efforts [15, 16].

How is the UK building a pro-cleantech environment? By coordinating its top-down efforts to support innovation, from development to market. For example, the UK's Low Carbon Innovation Coordination Group (LCICG) brings together public sector backed organizations who invest in clean energy innovation. By working together, these organizations leverage each other's expertise and seek to reduce wasteful duplication of efforts [17]. For anyone active in a highly regulated market such as energy, the experience in the UK shows how important it is to have government-backed innovation and R&D policies that support innovative technologies all the way to the point where they are sufficiently proven in the market—and to then provide market based policy mechanisms where needed [18]. This however without crowding out private initiative and entrepreneurship.

Next to sector-specific enabling conditions, a government's general stance toward startup and incubation activity also matters: many ideas that sound great in the lab will never become a world-leading product. To turn out a critical number of game changers, it is therefore helpful to benefit from startup pipeline buildup. This in turn is contingent not just on general attitudes toward entrepreneurship and risk-taking as is often mentioned. Rather, specific conditions such as access to capital and bankruptcy and tax law, as well as availability of information on players in a country's startup ecosystem condition how many ideas in the lab will eventually become investable opportunities. Italy's enabling startup legislation provides a fascinating case in point [19, 20].

10.4 Move from Patchwork to Transition

Patchwork of a couple of exciting solutions does not do the trick for an energy transition to succeed. Bottom-up innovation needs to encounter an enabling market environment that is created by top-down regulation. Germany is a remarkable example of how the demands of a grassroots movement managed to translate into a policy that came to be supported by all major political parties twenty years later. This offers insights into the mainstreaming of innovative ideas. After the first oil shock in 1973–74, the green movement kicked off and linked up with the anti-nuclear and peace movements. Concerns about environmental protection especially concerning the damage from acid rain, paired with the anti-nuclear movement based on fear of a nuclear catastrophe set free enormous creativity to look for alternative solutions. A growing majority of the population was demanding solutions; this led many scientists and politicians to look for a serious alternative to non-regenerative energy. Moreover, the green movement initially led to the emergence of a new political party, the Green Party. In other words, a grassroots movement led to the creation of an institutional infrastructure and presence in the parliamentary process that would relentlessly position the clean energy topic. By 1998, this former fringe party would find itself in government.

Once in power, this led to legislation that would guarantee producers of renewable energy a stable price (EEG). Rather than financing investments, a fixed price level was set so that investment made economic sense. The cost was rolled over to all consumers of electricity as a part of the price of electricity. This allowed for the establishment of a market and became the blueprint for other countries. The combination of strict legislation and a price signal made the German renewable energy market a place for solutions to emerge that would be competitive on the global market. German Chancellor Angela Merkel eventually even pressed ahead with a renewable energy transition without nuclear power. She initiated a definitive exit from nuclear energy after the Fukushima disaster.

Today, the German Energy Transition is still a work in progress. The renewable technologies on stream do not yet do the trick. With imports of 41 billion cubic meters (bcm) of natural gas from Russia (2013), energy dependence remains massive [21]. Moreover, German consumers currently face among the highest electricity costs in Europe, half of which consist of taxes and fees—seen that the country's energy-hungry industries are largely exempted from carrying the cost of the energy transition [22]. Notwithstanding, Germany represents the most advanced example of a national energy transition under way, inspiring many countries around the world, and offering valuable lessons on the way to the 80–100 % renewable economy. If some of the dynamic effects and technological advances work out, the energy transition could well become a new source of comparative advantage for German industry in due course.

To get a sense of the orders of magnitude involved, just consider the storage capacity coming on stream if Germany were not to miss its goal of putting one million electric cars on its roads by 2020. This with Vehicle-to-grid (V2G) capability and if swarm battery management—where batteries release electricity to the grid when it is needed and get charged when there is excess supply of electricity—becomes fully viable in scale on time. Even if these 2020 cars still only had today's BMW i3 125 KW batteries, and in the scheme 100,000 cars released electricity to the grid at any point in time, this would add up to 109,500 GWh in a year's 8760 h [23]. Under remote swarm management, electricity could be stored when there is excess supply of renewables in the grid, and discharged when there is net or peak demand. Now compare this to the next-generation French nuclear power plants of the 1650 MWe class in the European Pressurised Reactor (EPR) design [24]. If the nuclear power plant runs 80 % of the time, so 7008 of 8760 h in a year, it produces 11,563 GWh. In other words, provided there are sufficient solar panels and wind farms to retrieve the energy in the first place and enough electric cars to store and release it, a large-scale swarm battery system in the electric vehicles could enable capacity that substitutes for about ten nuclear power plants. Such flexible storage would make it much easier to stabilize the grid, allowing the percentage of renewables in the energy mix to go higher. We would get closer to turning field mice into 800-pound gorillas.

References

1. BP (2015) BP energy outlook 2035. London, UK. http://www.bp.com/en/global/corporate/press/press-releases/energy-outlook-2035.html. Accessed 2 Mar 2015
2. Graeber DJ (2015) New Zealand breaks renewable energy record. In: Energy Daily. http://www.energy.daily.com/reports/New_Zealand_breaks_renewable_energy_record_999.html. Accessed 2 Apr 2015
3. IEA (2012) World energy outlook 2012. http://www.worldenergyoutlook.org/media/weowebsite/2012/WEO2012_Renewables.pdf. Accessed 2 Mar 2015
4. The Parliament of the Commonwealth of Australia (2006) Australia's uranium—greenhouse friendly fuel for an energy hungry world. http://www.iaea.org/inis/collection/NCLCollectionStore/_Public/38/115/38115210.pdf. Accessed 7 Mar 2015
5. Tortajada C (2014) Dams: an essential component of development. http://www.thirdworldcentre.org/damsdevelopment.pdf. Accessed 9 Mar 2015
6. IEA (2013) Medium-term market report 2013. http://www.iea.org/textbase/npsum/mtrenew2013sum.pdf. Accessed 22 Mar 2013
7. IEA (2012) World energy outlook 2012. http://www.worldenergyoutlook.org/media/weowebsite/2012/WEO2012_Renewables.pdf. Accessed 22 Mar 2015
8. IEA (2013) Technology roadmap: wind energy. http://www.iea.org/publications/freepublications/publication/Wind_2013_Roadmap.pdf. Accessed 3 Mar 2015
9. IEA (2015) Geothermal power. http://www.iea.org/topics/renewables/subtopics/geothermal. Accessed 6 Mar 2015
10. IEA (2012) Renewable energy essentials: geothermal. http://www.iea.org/publications/freepublications/publication/Geothermal_Essentials.pdf. Accessed 2 Mar 2015

11. IEA (2013) Medium-term market report 2013. http://www.iea.org/textbase/npsum/mtrenew2013sum.pdf. Accessed 2 Mar 2015
12. Committee on Climate Change (2010) Building a low-carbon economy—the UK's innovation challenge. http://decarboni.se/sites/default/files/publications/137878/Building-low-carbon-economy-UKs-innovation-challenge.pdf. Accessed 17 Mar 2015
13. Low Carbon Innovation Co-ordination Group (2014) Coordinating low carbon technology innovation support: the LCICG's strategic framework. http://www.lowcarboninnovation.co.uk/working_together/strategic_framework/overview. Accessed 17 Mar 2015
14. Committee on Climate Change (2010) Building a low-carbon economy—the UK's innovation challenge. http://decarboni.se/sites/default/files/publications/137878/Building-low-carbon-economy-UKs-innovation-challenge.pdf. Accessed 17 Mar 2015
15. Beckmann K (2015) Meet the world's number 1 R&D player in sustainable energy: the Chinese Academy of Sciences. In: Energy post. http://www.energypost.eu/meet-worlds-number-1-academic-player-sustainable-energy-chinese-academy-sciences. Accessed 17 Mar 2015
16. International Energy Agency (2012) Energy technology perspectives 2012—pathways to a clean energy system. http://www.iea.org/etp/etp2012. Accessed 17 Mar 2015
17. Low Carbon Innovation Co-ordination Group (2014) Coordinating low carbon technology innovation support: the LCICG's strategic framework. http://www.lowcarboninnovation.co.uk/working_together/strategic_framework/overview. Accessed 17 Mar 2015
18. Committee on Climate Change (2010) Building a low-carbon economy—the UK's innovation challenge. http://decarboni.se/sites/default/files/publications/137878/Building-low-carbon-economy-UKs-innovation-challenge.pdf. Accessed 17 Mar 2015
19. Camere di Commercio d'Italia (n. d.) Qui si parla di imprese che fanno innovazione. http://startup.registroimprese.it. Accessed 17 Mar 2015
20. Italia Startup (2013) Restart, Italia. http://www.italiastartup.it/risorse/decreto-sviluppo-e-rapporti-dal-ministro-dello-sviluppo-economico. Accessed 17 Mar 2015
21. Oxford Institute for Energy Studies (2014) Reducing European dependence on Russian gas. http://www.oxfordenergy.org/wpcms/wp-content/uploads/2014/10/NG-92.pdf. Accessed 2 Apr 2015
22. Deutsche Welle (2014) German electricity price is half taxes and fees. http://www.dw.de/german-electricity-price-is-half-taxes-and-fees/a-17849142. Accessed 2 Apr 2015
23. BMW (2015) Technische Daten: BMW i3. http://www.bmw.com/com/de/newvehicles/i/i3/2013/showroom/technical_data.html. Accessed 16 Jul 2015
24. Areva (2015) EPR™ reactor: the very high power reactor (1650 MWe). http://www.areva.com/EN/global-offer-419/epr-reactor-one-of-the-most-powerful-in-the-world.html. Accessed 16 Jul 2015

Seven Game Changers for Our Energy Future

11

11.1 Take Storage from Linear to Algorithmic Progress

If we cannot solve the energy dilemma, there will be no convergence upon an economy that is sustainable in the long term, prosperous, and ever more circular—the Impact Economy. Success is a must. While it is never be good to be scared and passive, one would have to consider that the overshoot and collapse scenario will kick in eventually. So we better get it right. Let's look at the prospects of solar energy first. Solar power is the only alternative energy technology to date that has the scale potential needed to deliver the amounts of energy a growing and Westernizing global population needs. Moreover, in the best of locations, solar radiation delivers as much as 1.1 kW per m^2—think of the proverbial installation of solar panels in the Sahara [1]. This is relevant, because our modern lifestyle also imposes certain energy performance and availability requirements that we cannot just wish away. Nor can we mandate large-scale behavioral and civilizational change. This means that any breakthrough in energy will ultimately be associated with the ability to release and store power at a speed and energy density that is useful.

All devices or physical media that can store energy to perform useful processes at a later point in time, including chemical energy, electricity, gravitational energy, temperature differences and latent heat as well as kinetic energy are potentially relevant to step up progress in storage. Economics, performance, and safety are key, as well as whether energy can be stored only short-term or also longer term, such as between opposing seasons for heating (Fig. 11.1).

Rather than busying us with the many possible niche applications of alternative energy, it makes sense to generally consider a renewable energy's potential by comparing it with the energy density and peak power of well-established fossil fuels against which it must compete. In visual terms, consider a ragone chart, plotting the peak power of various energy sources against their specific energy (Fig. 11.2). The chart compares the performance of various energy-storing devices by plotting energy density (in Wh/kg) against power density (in W/kg) on a logarithmic scale to compare sources with very high power with limited power.

M. Martin, *Building the Impact Economy*,
DOI 10.1007/978-3-319-25604-7_11

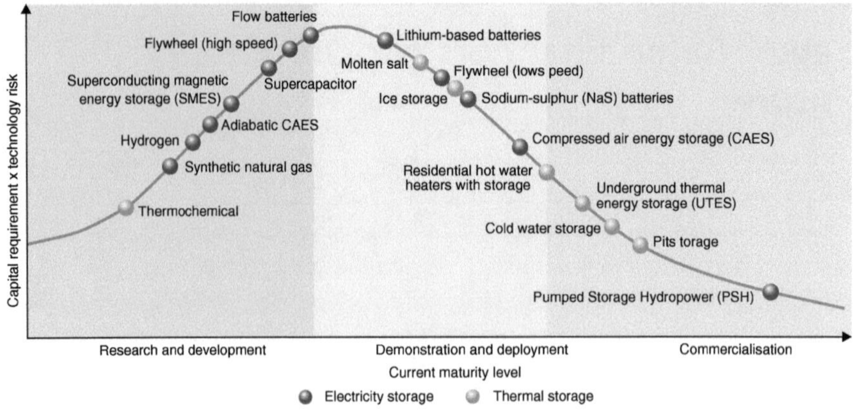

Fig. 11.1 Maturity of energy storage technologies (*source* Decourt and Debarre [2]) (Figure courtesy of Benoît Decourt)

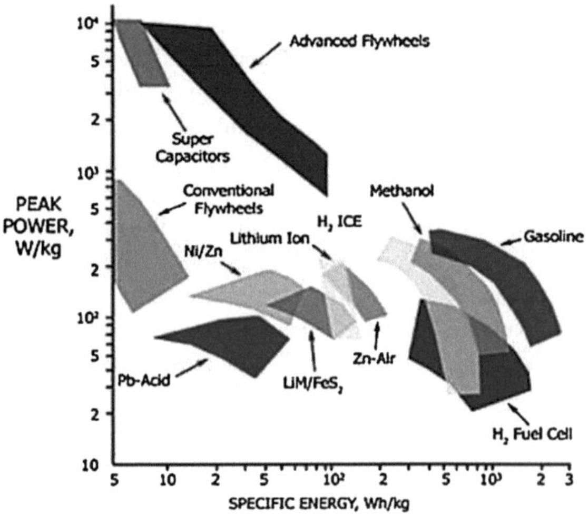

Fig. 11.2 Ragone plot of energy storage (*source* Ghoniem [3]) (Figure courtesy of Ahmed F. Ghoniem)

Note that while at this point only a theoretical solution, the LENR mentioned earlier would provide the kind of performance that would really help. The “E-Cat” would in fact be off the chart. It is theorized to have roughly four orders of magnitude more specific energy and three orders magnitude greater peak power than gasoline. If something like this ever hits the market, its impact will be massive.

It is time to ask more generally what we can do so that progress in storage starts to behave more like progress in the world of semiconductors as predicted by Moore’s Law with respect to the number of transistors in a dense integrated circuit, doubling every second year in the number of components per integrated circuit. The way forward is a combination of investing in innovations with game changing potential, and creating programs that can fast-track market entry.

11.2 Invest to Take Solar to the Next Level

As is well known, the potential of solar energy is nothing less than amazing. The amount of solar energy reaching the Earth has the potential to cover all our energy needs many times over; the solar energy the earth receives in one hour is enough to cover the world's energy consumption for an entire year [4]. However, we need to consider additional technologies to engage in solar harvesting: single-junction silicon-crystal photovoltaic cells that excite electrons across a band gap to produce electric power dominate the market. The best commercial silicon modules reach efficiencies just above 20 % [4]. According to the International Renewable Energy Agency (IRENA) Renewable Energy Power Costs Summary of 2014, photovoltaic LCOE prices are in the range of 0.12–0.20 USD/kWh, which is too expensive to be competitive with the fossil fuel LCOE of 0.05–0.1 USD/kWh [5].

There are several logical responses; all warrant an open eye for potential breakthroughs. Strategies to reduce LCOE prices generally pursue either (1) an increase in efficiency at the expense of higher cost per area, or (2) low cost solutions that can cover a large area, usually sacrificing efficiency. Moreover, there are additional technologies to consider that rely on semiconductors.

First, multi-junction cells which rely on several band gaps to harvest energy from different wavelengths of electromagnetic radiation are much more energy efficient (the current record in the lab stands at 46 % efficiency) [6]. However, they are still too costly to produce, with estimates indicating costs per unit of power produced almost being double that of single-junction cells (let alone fossil fuels) [7]. This renders inventions that dramatically reduce cost particularly exciting.

A second logical avenue are thin-film solar cells that rely on band gaps to produce electricity, but do not use expensive silicon based semiconductors. Issues to resolve in thin-film solar are (1) low efficiencies and (2) the toxicity of thin-film materials such as cadmium telluride (CdTe), a stable crystalline compound formed from cadmium and tellurium, or copper indium gallium diselenide ([$Cu(In_xGa_{1-x})Se_2$], abbreviated CIGS).

Finally, given the inherent large potential of solar energy and technology risk, the key answer to the *how-to* question is exploring completely different approaches to solar harvesting that are now becoming a practical possibility because of current advances in nanotechnology. These include (Table 11.1):

- Perovskite solar cells;
- Organic solar cells;
- Quantum dot cells; and
- Plasmonic cells.

Table 11.1 Performance and readiness of solar cell technologies (*source* author) [8]

Technology	Description	Best research efficiency (%)	Cost *USD/W*	Typical length of warranty	Readiness	Environmental impact
Monocrystalline	Oldest cell technology and most widely used	25	~0.70	25 years	Mature	High level of embedded energy to produce crystalline silicon
Polycrystalline	Less silicon waste in production process	20.40	~0.60	25 years	Mature	High level of embedded energy to produce crystalline silicon
Amorphous Si	Tend to degrade faster than crystalline-based panels; Low availability on the market	13.40	~0.50	10–25 years	Medium	Medium
CdTe	Tend to degrade faster than crystalline-based panels; Low availability on the market	21.50	~0.70	10–25 years	Medium	Toxicity of cadmium
CIS/CIGS	Tend to degrade faster than crystalline-based panels; Low availability on the market	21.70	~0.70	10–25 years	Medium	Medium
Dye-sensitized	Must reduce cost and increase durability for commercialization	11.90			R&D	Low impact
Organic	Must reduce cost and increase durability for commercialization	11.10			R&D	Low impact
Perovskite	Must reduce cost and increase durability for commercialization	20.10			R&D	Low impact
Quantum dot	Must reduce cost and increase durability for commercialization	9.90			R&D	Low impact
Multi-junction CPV	Very expensive and used when cost is less of a factor (e.g.: space, CPV)	46			R&D	Low impact

There is good reason to be especially excited about the application of advances in quantum physics to the field of solar energy.

For example, in quantum physics, particles are subject to different constraints than in classical physics because their dual properties include both a particle and a wave. This means that electron tunneling could be used for electricity generation, rather than only the well-established band gap mechanism from solar photovoltaics. This is the core idea behind the so-called "plasmonic effect," a phenomenon where an electromagnetic radiation (e.g. from the sunlight) can induce plasma oscillations in a metal and produce electric current [9].

For example, applying quantum physics and nanotechnology to come up with a next-generation solar cell, US company SciTech Solar is developing a completely new way of harvesting solar energy without the band gap limitations of conventional solar cells. Their single element solar cell utilizes rectenna technology, which essentially consists of an antenna and a diode combination. The nano scale antenna captures sunlight at a high efficiency and the diode then rectifies the energy into usable direct current.

As the antennae can be tuned to respond to any frequency of solar radiation, the solution can utilize a wide spectrum of the sun's energy. This makes it possible to achieve efficiencies of well over 50 %, perhaps even 70 %. Moreover, such cells would be cheap to produce and would not require the use of toxic materials in the production process [10].

11.3 Not Sexy, but Easy: Run with Energy Efficiency

Service and residential buildings are currently responsible for 40 % of energy consumption in Europe, ahead of transportation and the industrial sector [11]. Energy efficiency can only provide a partial answer to the energy crisis. Notwithstanding, this makes reduction of primary energy demand via greater energy efficiency a natural candidate for investments into fast tracking the energy transition. Consider for example the potential of insulation solutions that offer up to ten times higher insulation performance compared to conventional insulations, are space saving, and reduce energy costs.

The International Energy Agency (IEA) defines energy efficiency as a way of managing the growth in energy consumption [12]. A device can then be defined as more energy efficient if it delivers (1) more services for the same energy input, or (2) the same services for less energy input [12]. By using energy more efficiently, we can lower our energy bills, reduce our reliance on external suppliers of oil and gas and help protect the environment. Many relevant initiatives are under way. For example, the EU has set itself a 20 % energy savings target by 2020 when compared to the use of energy in 2000 [13]. This is roughly equivalent to turning off 400 power stations [14]. Moreover, British Gas research has found that nine out of ten smart meter customers are already taking simple daily steps to manage their home energy use. Nearly two-thirds of smart meter customers have identified savings estimated up to GBP 75 per year [15]. By the end of 2020, over 50 million smart meters are projected to be fitted in over 26 million households across Wales, Scotland and England [15].

The private sector includes a great deal of potential as well. Companies worldwide spend an average 40 % more energy due to inefficiencies [16]. To reduce demand without negatively affecting the operations of their business, firms need a centralized solution to analyze energy consumption, to detect inefficiencies, and automatically solve them. As the information revolution gives birth to the "Internet of Energy," the solution includes next generation gas and electricity meters that are smart and capable of showing the user via an interactive web portal how much energy they are using, how much that costs, and how to better manage personal or company energy use.

Can such energy efficiency solutions break new ground? Yes, if they can dramatically expand reach and depth: the next step forward here does not lie in the technology itself, but in designing technologies that dramatically raise a solution's ability to scale. There are several pathways to achieve this, including lowering cost and enhancing convenience. Given the interlinkage of information technology and energy technology, such solutions can become complex very fast.

For example, Spanish company OpenDomo provides state-of-the-art smart meters at roughly half the price of its competitors [17]. Unlike most other smart meter providers, these meters have an integrated gateway that is capable of sending information directly to the OpenDomo cloud or any other server using a standard Ethernet cable or Wi-Fi. This reduces complexity and system setup cost.

Beyond measuring energy, gas and water usage, the smart meter devices are also capable of measuring temperature, humidity or pH acidity, among others. In order to eliminate any type of inefficiency, the device is then also capable of automation and remote control of facilities via a cloud. OpenDomo is currently using machine learning and data mining techniques to develop a high impact software module that can integrate artificial intelligence into the firm's energy management platform.

11.4 Take E-Mobility beyond Entertainment

In 2014, the global auto industry expected to have sold 85 million vehicles, up from 82 million in 2013, with sales forecast to break 100 million in 2018 [18]. In comparison, Tesla sold its flagship Model S 20,000 times in 2013 [19]. A 13 % of global greenhouse gas emissions come from transportation that is currently mainly powered by petroleum-based fuels (gasoline and diesel), which are burned for air, marine, rail and road transportation [20]. Whether German carmaker Volkswagen's 2015 "clean diesel" scandal will ultimately accelerate the large-scale adoption of electric vehicles remains to be seen.

To locate solutions for this sector, electro mobility (or "e-mobility") solutions that enable the electric propulsion of vehicles and fleets are especially exciting. This comes down to the combined use of electric powertrain technologies, in-vehicle information, and communication technologies as well as dedicated infrastructure for recharging.

When most people hear of electric vehicles, they think of one word, batteries. However, batteries are not the only energy source option for renewable mobility. Hydrogen is proving itself a promising clean alternative to fossil fuels that not only competes with, but also complements batteries. Which technological pathway will win is not yet clear, so we need to consider full electric vehicles as well as plug-in hybrids and hydrogen fuel cell vehicles that convert hydrogen into electricity. Let's be clear though: until the storage problem is solved, e-mobility will remain a trophy exercise for carmakers pleasing regulators, some governments, and lifestyle customers. Serious progress on the mobility front is urgently needed and swarm battery management in electric vehicles could provide a major boost to stabilizing the grid.

Hydrogen as a fuel is very interesting, in spite of the many challenges still to be solved. By weight, hydrogen contains nearly three times the energy of petrol [21]. However, in its gaseous state, hydrogen has a volumetric energy density that is inferior to hydrocarbons [22]. To compensate for this, the industry standard for fuel cell electric vehicles (FCEV) such as the Toyota Mirai consists of compressing hydrogen to 700 times atmospheric pressure and then storing it in carbon fiber reinforced cylindrical tanks. The tanks must be cylindrical because this form is strongest in resisting high pressure—but this limits design freedom [23]. There are many other drawbacks to compressed hydrogen gas tanks including high pressure, weight, volume, adaptability and cost. A 2012 report by Strategic Analysis Inc. on behalf of the US Department of Energy found that a 700 bar hydrogen storage system could cost above USD 20/kWh of capacity. Compressing hydrogen is also an energy intensive process, resulting in losses of 5–15 % of the total energy. The key obstacles to overcome for hydrogen to become a major clean energy sources are storage and CO_2-free production.

In France, we found a company with a potentially groundbreaking solution to both problems: if hydrogen is hard to store, why not produce it on demand? HySilabs has developed a catalyst that can release hydrogen from water on demand in the lab, and at atmospheric pressure. This technology is groundbreaking in the sense that it removes the need to store hydrogen entirely. All you need is water and a non-hazardous catalyst mixture. With the same amount of liquid contained in a normal petrol tank, enough energy can be stored to offer the same driving range as fossil fuels, but with the only emission being harmless H_2O. The applications go far beyond vehicles. For example, this technology could be used to store and supply energy to small electronics, power tools, and isolated power generators, and address both the storage and the CO_2-free production problem.

11.5 Carefully Pick Winners in the Infrastructure Buildup

At the turn of the millennium, experts predicted that wind power would globally reach 30 GW in 2010—in that year, capacity reached 200 GW, and today it stands at 370 GW globally [24]. Similarly, in 2002, it was predicted that solar power would add 1 GW per year by 2010—yet in 2010, 17 GW of solar capacity were added globally, and in 2014, 48 GW [25]. Put differently, we are in the early days of an incredible capacity build up in the renewables industry. In the near future, people will speak of "Big Solar" and "Big Wind" in similar ways as we do today of "Big Oil" and "Big Pharma." Former US Vice President Al Gore is on his way to becoming the world's most widely known climate billionaire. In short, the renewable energy space is not only important for our future, but also big business for VC and later stage investors.

Market forces everywhere are engaged in building parts of the emerging alternative energy infrastructure. This leads to interesting investment and impact opportunities, often connecting the Internet of Energy to brick and mortar businesses. Notwithstanding, picking winners is difficult at this early stage of market development. For example, the Italian startup Electric Drive Italia (EDI) supplies its customers with a turnkey and tailor made e-mobility integrated system having its core in the ITC platform called B.O.M.T. (Banking-Operation-Maintenance-Telematics). The B.O.M.T. platform assures banking, operational, maintenance and advanced telematics systems, representing the guarantee of the project business plans that are agreed in advance with the single customers. The system is made of e-cars/buses, charging stations, IT systems for charging station control, car booking management and fleet management. The firm's customers are public and private organizations in the semi-public transportation area. The goal of their service offering is to render e-mobility easy by offering a simple and cheap e-drive experience that solves a critical problem affecting its convenience: fast and safe battery charging. A whole host of ancillary services is conceivable, ranging from fleet management to maintenance and user telematics as well as banking services. Cars could even become part of an energy trading swarm where one can buy electricity from the local utility company at a lower or negative price in the case of an excess supply and then sell it on later. A medium sized Italian city alone would require at least 20 charging stations: one can only imagine the tsunami of capital investment in infrastructure once e-mobility is fully under way!

11.6 Digital Is Yin—Hardware Is Yang

One of the most fascinating observations from systematically scouting for breakthrough clean energy solutions is the number of "IT plays" out there. For sure, digital transforms all value chains. It's the "yin" of the energy transition. To stay in the previous e-mobility area for a second, just consider the potential of online applications to find the next EV charging stations, or the potential to use electric vehicles as an orchestrated swarm of batteries, to be charged when electricity is cheap and abundant, and to be released when peak demand hits, discussed earlier. A host of residential and commercial applications that optimize energy use, e.g. by remotely dimming LED traffic lights as needed, is currently being conceived and rolled out to optimize how we use energy.

Though important and exciting, that alone will be insufficient to solve the energy dilemma. The "yang" is equally needed. Today's economy still benefits from inventions made more than fifty years ago such as the transistor or nanotechnology;

it now seems that most of our talents and US West Coast VCs are focused mainly on information technology applications rather than also giving a hand to the hardware breakthroughs that are badly needed. Information technology is bound to open up amazing new possibilities; but in and by itself it will not be able to enable us to succeed at solving the energy problem. Barriers to entry are low for IT—after all, this is about programming some app—and historically valuations for successful IT startups have been such that one may wonder why young talents go into other fields at all. The relevant *how-to* conclusion is: we need to find ways not to shower all the innovation funding on digital solutions; rather, design appropriate pathways to take forward hardware as well!

11.7 Keep "Out-of-the-Box" Thinkers on the Radar

Groupthink and conformity dominate most companies and organizations. However, this is not how progress is achieved. To secure any kind of breakthrough, open-mindedness and creativity are key. To end up in the "our-future-yea" camp, we will need a series of major breakthroughs. So how can we raise the likelihood to get it right?

A lot more conventional work is needed for sure. Notwithstanding, one needs to be mindful that progress sometimes happens in unexpected ways—and maybe even during holidays. Take an example from another field: Sir Alexander Fleming, a UK successor to the iconic French founding father of microbiology Louis Pasteur. Searching for anti-bacterial agents for ten years, he discovered penicillin in 1928. Departing for a family holiday, he left his lab untidy. One culture of staphylococci became contaminated with a fungus while he was vacationing and destroyed the other colonies surrounding it. Fleming was open-minded enough to recognize the extraordinary event; the rest is history.

The next Flemings are the ones who can upset the apple cart—so we scouted for them. A fascinating part of the search was the number of people who consider themselves contrarians or may even be active conspiracy theorists. They argue that they have the breakthrough solution but could not show it to us (or to anyone else for that matter), either because the world is not ripe, or because they are afraid. Others argue that the likes of Nikola Tesla had already worked out groundbreaking solutions, but that their inventions had been suppressed. From a scouting perspective, the moment of truth comes when the prototype is or is not presented. Moreover, assessing technological readiness is generally a good yardstick to anticipate the steps to master for a working prototype to become a viable product. Notwithstanding, given the need to locate out-of-the-box solutions, keeping one's ear to the ground and looking for the unconventional may just hold a missing element for the energy transition to succeed. We need to accept that different pathways are needed to productively engage with a lone inventor than with a member of the science establishment.

References

1. Newport (2015) Introduction to solar radiation. http://www.newport.com/Introduction-to-Solar-Radiation/411919/1033/content.aspx. Accessed 30 Mar 2015
2. Decourt B, Debarre R (2013) Electricity storage. In: Factbook. Schlumberger Business Consulting Energy Institute, Paris, and IEA (2015) Technology roadmap energy storage. http://www.iea.org/publications/freepublications/publication/TechnologyRoadmapEnergystorage.pdf. Accessed 22 Mar 2015
3. Ghoniem AF (2011) Needs, resources and climate change: clean efficient conversion, fig 38, p 49. Prog Energy Combust Sci 37:15–51. http://ac.els-cdn.com/S0360128510000341/1-s2.0-S0360128510000341-main.pdf?_tid=f7c4a6e2-4727-11e5-81cc-00000aacb35e&acdnat=1440067449_9c692e22849e82c9e66fb86f0ce5903b. Accessed 19 Aug 2015
4. Shwartz M (2015) Perovskites provide big boost to silicon solar cells, Stanford study finds. Stanford Precourt Institute for Energy. https://energy.stanford.edu/news/perovskites-provide-big-boost-silicon-solar-cells-stanford-study-finds. Accessed 11 Mar 2015
5. IRENA (2014) Renewable energy power costs of 2014 executive summary. http://www.irena.org/DocumentDownloads/Publications/IRENA_RE_Power_Costs_Summary.pdf. Accessed 3 Mar 2015
6. Colthorpe A (2014) Soitec-Fraunhofer ISE multi-junction CPV cell hits world record 46% conversion efficiency. http://www.pv-tech.org/news/soitec_fraunhofer_ise_multi_junction_cpv_cell_hits_world_record_46_conversi. Accessed 5 Mar 2015
7. Fraas L, Partain L (2010) Solar cells and their applications. Wiley, New Jersey
8. Martin M (2015) Energy transition fast forward! Scouting the solutions for the 80–100 % renewable economy, p 30. Impact Economy. http://www.impacteconomy.com/papers/IE_PRIMER_APRIL2015_EN.pdf
9. Dmitruk N, Korovin A (2013) Plasmonic photovoltaics: near-field of a metal nanowire array on the interface for solar cell efficiency enhancement. In: Semiconductor Science and Technology. http://iopscience.iop.org/0268-1242/28/5/055013. Accessed 6 Mar 2015
10. SciTech Solar (2011) Technology overview. http://www.scitechsolar.com/technology.html. Accessed 3 Mar 2015
11. EEA (2013) Final energy consumption by sector (CSI 027/ENER 016). http://www.eea.europa.eu/data-and-maps/indicators/final-energy-consumption-by-sector-5/assessment. Accessed 6 Mar 2015
12. International Energy Agency (2015) Energy efficiency. http://www.iea.org/topics/energyefficiency. Accessed 6 Mar 2015
13. Wesselink B et al (2010) Energy savings 2020: how to triple the impact of energy saving policies in Europe. http://www.roadmap2050.eu/attachments/files/EnergySavings2020-FullReport.pdf. Accessed 7 Mar 2015
14. European Commission (2015) Energy efficiency: saving energy, saving money. http://ec.europa.eu/energy/en/topics/energy-efficiency. Accessed 7 Mar 2015
15. British Gas (2015) Smart meters: everything you need to know. In: The Guardian. http://www.theguardian.com/british-gas-smart-meter-challenge/2014/aug/26/smart-meters-everything-you-need-to-know. Accessed 7 Mar 2015
16. International Monetary Fund (2013) Energy subsidy reform: lessons and implications. http://www.imf.org/external/np/pp/eng/2013/012813.pdf. Accessed 7 Mar 2015
17. OpenDomo (2015) The OpenDomo register. http://www.opendomo.com. Accessed 7 Mar 2015
18. Ramsey M (2013) Global car sales seen reaching 85 mln in 2014: HIS. http://www.marketwatch.com/story/global-car-sales-seen-reaching-85-mln-in-2014-ihs-2013-12-16. Accessed 7 Mar 2015
19. Shahan Z (2014) Global sales of Tesla model S may have surpassed 25,000. http://cleantechnica.com/2014/01/09/global-sales-tesla-model-s-surpass-25000-2013/. Accessed 7 Mar 2015

20. Greene DL et al (2011) Reducing greenhouse gas emissions from U.S. transportation. Center for Climate and Energy Solutions. http://www.c2es.org/docUploads/reducing-transportation-ghg.pdf. Accessed 8 Mar 2015
21. Office of Energy Efficiency & Renewable Energy, NA (2015) Hydrogen storage—Basics. http://energy.gov/eere/fuelcells/hydrogen-storage-basics-0. Accessed 12 Mar 2015
22. The SBC Energy Institute (2014) Hydrogen-based energy conversion—More than storage: system flexibility. http://www.sbc.slb.com/~/media/Files/SBC%20Energy%20Institute/SBC%20Energy%20Institute_Hydrogen-based%20energy%20conversion_FactBook-vf.ashx. Accessed 11 Mar 2015
23. Toyota (2015) Powering the future—Hydrogen fuel cell vehicles could change mobility forever. http://www.toyota-global.com/innovation/environmental_technology/fuelcell_vehicle/. Accessed 12 Mar 2015
24. OffshoreWIND (2015) WWEA: almost 370 GW of wind power installed globally. http://www.offshorewind.biz/2015/02/09/wwea-almost-370-gw-of-wind-power-installed-globally. Accessed 12 Mar 2015
25. Lins C et al (2014) The first decade: 2004—2014: 10 years of renewable energy progress. In: Renewable Energy Policy Network for the 21st Century. http://www.ren21.net/Portals/0/documents/activities/Topical%20Reports/REN21_10yr.pdf. Accessed 12 Mar 2015

Time to Aim High—Use Our Talents and Instruments

12

12.1 A Tale of Two Programs: Climate KIC and ARPA-E

If it's time to aim high, what about the larger *how-to* question on solving the energy dilemma, next to the specific actions in the key technology theaters outlined in the previous chapter? The EU and the US both have dedicated initiatives under way to help surface and support the scientific research needed to take technologies to market that are relevant to the energy transition. They are taking different routes to source renewable energy innovations. In isolation, they will be insufficient; in conjunction they can be powerful.

Within the framework of its Lisbon Strategy for Growth and Jobs, the EU aims to create a green and innovative economy. To provide a dedicated platform for these efforts, the EU established a flagship institute headquartered in Hungary to integrate innovation, research and growth in 2008, the European Institute of Innovation and Technology (EIT). The goal was to help overcome Europe's perceived innovation gap [1]. EIT runs two main initiatives relevant to the energy field, namely the Knowledge Innovation Communities ("KIC") Climate KIC and KIC InnoEnergy. KIC InnoEnergy focuses on locating new technologies and products to enable sustainable energy for a climate-neutral Europe [2]. Climate KIC's brief focuses more broadly on sustainable cities, adaptive water management and carbon neutral production systems [3].

To date, Climate KIC has financially supported over 175 startups via two funding platforms, the Climate KIC Accelerator and the Climate Launchpad [3, 4]. The focus is on very early stage support that is not capital intensive. The accelerator program provides staged support to breakthrough ideas related to a new technology with a substantial climate impact. In the initial "fundamentals" stage, grantees receive up to EUR 20,000 in funding as well as coaching; in the subsequent "validation" stage, applicants meet a minimum of 50 potential clients and can access another EUR 25,000 in funding; in the final "delivery" stage, applicants secure a first customer and receive up to EUR 50,000 in funding. The Launchpad program runs a cleantech business plan competition open to participants from over

M. Martin, *Building the Impact Economy*,
DOI 10.1007/978-3-319-25604-7_12

20 countries where winners can access EUR 10,000 in funding, as well as becoming eligible for the accelerator program. KIC InnoEnergy similarly funds startups via a variety of windows, including partnering with crowdfunding platforms [5]. Without going into the level of effectiveness of these programs today, it is apparent that significantly more resources need to be deployed to build up the innovators pipeline to a point where we can reasonably expect a whole swarm of gamechanging technologies to hit the market eventually.

The Advanced Research Projects Agency-Energy (ARPA-E), in the US takes a different approach. ARPA-E was modeled after the successful DARPA program of the U.S. Department of Defense. DARPA was responsible for bringing us the earliest predecessor to the Internet, global positioning systems (GPS) and many other technological breakthroughs. According to the agency's acting director, ARPA-E's mission is to "catalyze and accelerate energy technologies that will enhance the economic and energy security of the US" [6]. It does this by funding high-risk, high-reward research in technologies that are too early for the private sector and could revolutionize the way we generate, store and use energy [7]. ARPA-E was created in 2007 by the America COMPETES act, but did not officially come into being until 2009 when it was allocated USD 400 million in funding as a part of the American Recovery and Reinvestment Act, better known as the U.S. economic stimulus package [8]. In order to receive funding, applicants are initially invited to submit eight-page "concept papers" that describe the technical concept, followed by a full application if they pass the first cut. Submissions are only reviewed if they are deemed to be compliant, meeting eligibility requirements and conforming to content and form requirements set by the Funding Opportunity Announcement (FOA), and responsive; they also must fall within the technical parameters described in the FOA [8]. Submissions meeting these requirements are then assessed according to evaluation criteria specific to each FOA. ARPA-E states that in its considerations, it may also use program policy factors, which are "broad policy considerations determined by agency needs and priorities" [8]. Those who are selected receive assistance in the form of Cooperative Agreements or Technology Investment Agreements with awards ranging from USD 500,000 to USD 20 million [9]. ARPA-E then works closely with its projects for one to three years by establishing and monitoring project milestones and has substantial involvement in the management and direction of every project.

To date, ARPA-E grant recipients have applied for at least 34 patents and five projects have generated spinoff companies or launched Initial Public Offerings (IPOs) [10]. During 2011, in the midst of widespread program cuts, the U.S. Congress voted to increase ARPA-E's funding by 50 % [11]. In fact, ARPA-E has been one of the few programs that both Democrats and Republicans value. This is in stark contrast to the Department of Energy's Loan Guarantee Program, which is notorious for lending USD 527 million to the solar company Solyndra of California, which ended in a controversial bankruptcy [12]. In comparison, ARPA-E has been inexpensive with an annual budget less than the amount loaned to Solyndra.

The program is not without criticism however. David Victor of the University of California San Diego observes that, "the parts of the innovation chain that are the hardest in energy are the scaling up and demonstration phases, and that's precisely what ARPA-E is not designed to do" [13]. In other words, ARPA-E is not designed to help startups master the valley of death between initial launch and successful achievement of scale and financial viability mentioned earlier. ARPA-E's annual budget of USD 300 million is spread across dozens of projects. Critics argue this is insufficient for many energy technologies that may require hundreds of millions of dollars to build large-scale demonstrations before private investors are willing to move in [13]. This was precisely supposed to be the role of the above mentioned DOE Loan Guarantee Program which has come under fire after some high profile failures. Although there is much excitement surrounding ARPA-E, only time will tell of its innovation model for the energy ecosystem will yield results that are comparable to those of DARPA. To hedge our bets, we will need additional instruments in the precompetitive stage.

12.2 The How-to Formula Links Capital, Talent—and Daring

Forty years after environmental concerns first burst onto the scene, renewable energy is now an idea whose time has come. If you are skeptical, consider that even the large established players—who had for the most part kept dragging their feet—are waking up to this new reality. Decentralized wind and solar power is hitting customer demand and thus the bottom line. Incumbents need to consider how to incorporate renewables into their business models. They are asking themselves what mix of activity matches the company's current skills and its current geographic presence: in a world where energy provision will become more decentralized, how can one build market share in a business sector that is in formation, and where classical capital-intensive upstream exploration plays less of a role [14]?

Creative destruction is upon us; there is work ahead. The clock is ticking to get it right. Take E.ON, Germany's largest utility. On November 30, 2014, the company announced that it would split itself up into a renewables and a fossil business (colloquially referred to as "E.OFF"). In 2016, the company plans to spin off its fossil fuel and nuclear business in order to focus on its renewables, distribution and customer solutions [15]. After the German government initiated the transition to renewables in 2000, Johannes Teyssen, E.ON's chief executive, argues that, "the traditional business model for utilities has broken apart" [16]. E.ON's CEO further stated that in his view, technology and not politics is ultimately driving the transition away from traditional sprawling utilities. "In the end technology breaks the value chain in two pieces, and not fundamentally politics or regulation, he said. "The main driver is technology and customers" [16].

Driven by a combination of these factors, Germany is now generating nearly a quarter of its electricity from renewable sources [17]. The German government's goal is to boost this figure to 45 % by 2035 [17]. New technologies enabling clean distributed generation and intelligent demand side management are shifting the focus back to the customers and away from giant centralized utilities. E.ON is the first large European player to seek to adjust its structure. In this highly regulated industry, government will no doubt influence how exactly these plans will play out. Other energy companies around the world are bound to follow suit. Incumbents in oil, gas and utilities are wondering when they should move. Just like competitors such as Airbnb and Uber came out of nowhere in the hotel business and the taxi business, we can safely assume that there will be a couple of fresh competitors who will suddenly be on everyone's radar in the energy industry.

To win the race against time, the goal has to be to find and nurture them much faster. Incumbents need to figure out how to deal with new players who have embraced new renewable technologies and are hungry for a piece of the lucrative energy market. For society, the question is whether progress can be achieved fast enough, and investors ask where money can be made. Next to those who are originating the technological progress currently under way, private sector incumbents, governments and investors are all key to solving the energy puzzle. Defensive communication strategies look increasingly insufficient.

Given the intermittent nature of much of renewable energy production and prevailing energy consumption patterns in industrial and residential applications, as well as competition from fossil substitutes, clean energy will definitely have to be produced much more efficiently in the future. Moreover, energy storage systems in the 80–100 % renewable economy will need a much larger storage capability, the ability to very rapidly charge and discharge cycling, as well as improved endurance. To invent our way out of these (and other) challenges fast, we need capital, ideas, and talent. In addition, we need to be attentive to enabling technologies that have a ripple effect, that are game changers just like the invention of the wheel or the printing press were in their days. The wheel changed the game in 3500 BC. When people were already building canals and sailboats or casting metal alloys in the Bronze Age, they still had no carts. 5500 year later, we take cheap long-distance overland transport for granted. Without the wheel, it would have been impossible to get where we are. Similarly, now knowledge travels ever faster, since the invention first of the printing press in the 1430s and the onset of the communication revolution—kicked off when Samuel Morse invented the electric telegraph in 1836. This goes to show that today's utopia may be standard practice fifty years from now.

I have written this book because technology is ultimately also a question of imagination and daring. For illustration, take the Luna Ring concept developed by Japanese scientists [18]. The Luna Ring is a theorized belt of solar cells around the 11,000 km lunar equator that generates electric power—which it transmits to the Earth from the near side of the Moon, which always faces the Earth. To achieve this, the Luna Ring uses microwaves and high-powered laser beams for energy transmission. The Luna Ring could generate enough electricity to power the entire

Earth. 20 km-diameter antennas are envisioned to transmit power to the receiving rectennas and a high-energy-density laser to be beamed to the receiving facilities. Transmitted anywhere on the Earth, electricity could then be fed to the grid, stored or used to produce hydrogen. While fascinating, the Luna Ring is not an investable proposition today though. The Japanese scientists working on this project hope to start building the Luna Ring in 2035: after machines and equipment from the Earth are assembled in space and landed on the lunar surface for installation, robot-controlled solar production plants would then move automatically while producing solar cells from lunar resources, and installing them. Science fiction stuff indeed.

12.3 Cut Through Energy's Gordian Knot

For our energy imagination, not even space may be the limit. We need solutions here and now though. Whatever is rolled out needs to work and be cheap—otherwise large-scale adoption will remain a fantasy. For example, if a recent discovery makes it to production, organic batteries could become key to getting to the storage solutions we need. The price signal is good: MIT scientists have created an organic flow battery that costs only USD 27 per kWh compared to metal batteries at USD 700 per kWh. This is nearly a 97 % saving. As we chart a promising path toward our energy future, it is essential to keep things simple and practical whenever possible. For example, we do not know when low-energy nuclear reactions will translate into products, but a premier research institution should have this technology on the radar. In the meantime, many new technologies are viable today; they need thoughtful and ambitious venture capital to graduate from their existence in the lab.

To win the race against time, we need to cut through the Gordian knot of energy innovation. This requires intelligent action at the micro as well as macro levels. Discussing multibillion dollar figures and theoretical CO_2 emissions reduction targets alone is insufficient to get micro-level incentives right. Success can only be achieved collectively, be based on private initiative and on brilliance in science. It is rendered more likely if every actor intervenes where they have leverage. This means both putting up the resources and backing the talent who can invent and deliver.

First, next to the efforts under way at the level of the international community (cue: "climate deal"), governments around the world need more ARPA-E type facilities. They can strengthen renewable energy innovation in the precompetitive stage, and be combined with programs to remove obstacles from technologies' pathways to market, and facilitating innovative financing.

Second, private investors need to fully grasp the magnitude of the industrial transformation that is building up and the associated opportunities.

Third, inventors are getting more and more support. We have however not yet seen the winning formula how public sector professionals can best support the scientists that are most deserving, while paying attention to both their track record as well as preserving the flexibility to act fast and back true genius when it manifests itself.

Fourth, it is natural for the established energy industry to be worried about its bottom line. Notwithstanding, the paradigm change is approaching no matter what. Incumbents are well advised to develop business strategies to take economic advantage of the renewable energy revolution that is around the corner, rather than going down fighting it.

Finally, civil society and philanthropy have a productive role to play in connecting the generations and citizens. In our world, centrifugal forces could take over if we can neither provide energy to the energy poor at the Base of the Pyramid, nor prevent a new Migration Period once climate refugees start to cross borders on a massive scale. In due course, climate refugees and geostrategic shifts will become a problem not just for regions which people will leave, but also for scarcely populated resource-rich countries and regions such as Russia's Siberia or Argentina that look vulnerable to climate-driven inward migration, or shrinking populations such as in Western Europe. We had better come up with some answers.

Just like our best scientists dare every day, we also need to explore new avenues to accelerate innovation and investment in the renewable energy space. For those who get it right, many fresh opportunities to make both money and a positive contribution are around the corner. To solve the energy problem at the scale required, we need all the talent we can get. Without being naïve about people's capabilities and motivations, we need to find effective, sometimes new, ways of working together: scientists, investors, and business developers. Even adepts of conspiracy theories assuming that all the solutions needed already exist but have been suppressed may have a contribution to make. In the age of the citizen and the investor, it is up to all of us to find, nurture and back the next generation of clean energy solutions. We do not know how fast we will progress. However, as master conductor Herbert von Karajan reminds us, we can safely assume that it would be a mistake to aim too low [19]. There is too much at stake. It's time to aim high.

Thus far, we have looked at the *how-to* questions for big business, textile and garments (as a placeholder for long supply chains), and energy. It is now time to explore how we can invest in innovation on the scale that is required. This is the focus of Part IV. Fixing energy means fixing the mother of all markets. We cannot cut through this Gordian knot if we do not have the pathways to allocate the capital needed to back the investment in the technologies that will change the game. A further complication? Certainly, but this is about our future—yea or nay!

References

1. European Institute of Innovation and Technology (2015) EIT—Making innovation happen. http://eit.europa.eu. Accessed 21 Mar 2015
2. KIC InnoEnergy (2015) The KIC InnoEnergy register. http://www.kic-innoenergy.com. Accessed 21 Mar 2015
3. EIT Climate (2015) Welcome to Climate-KIC, the EU's main climate innovation initiative. http://www.climate-kic.org. Accessed 21 Mar 2015
4. Climate-KIC (2014) Business plan. http://www.ctt.upv.es/documentos/BP%202014%20-%20draft%20-%2026%2008%20no%20TC%20vz.pdf. Accessed 21 Mar 2015

5. KIC InnoEnergy (2015) The European company for innovation, business creation and education in sustainable energy. http://www.kic-innoenergy.com. Accessed 21 Mar 2015
6. DARPA (2015) History. http://www.darpa.mil/About/History/History.aspx. Accessed 21 Mar 2015
7. The U.S. Department of Energy (2013) Advanced Research Projects Agency-Energy - ARPA-E Strategic Vision 2013. http://arpa-e.energy.gov/sites/default/files/ARPA-E_Strategic_Vision_Report_101713_0.pdf. Accessed 21 Mar 2015
8. ARPA-E—The U.S. Department of Energy. http://arpa-e.energy.gov/?q=faq/general-questions. Accessed 21 Mar 2015
9. EERE Network News (2009) DOE launches the Advanced Research Projects Agency-Energy, or ARPA-E. http://apps1.eere.energy.gov/news/news_detail.cfm/news_id=12478. Accessed 21 Mar 2015
10. ARPA-E—The U.S. Department of Energy. http://arpa-e.energy.gov/?q=faq/general-questions. Accessed 21 Mar 2015
11. MIT Technology Review (2012) Can ARPA-E solve energy problems? http://www.technologyreview.com/news/427136/can-arpa-e-solve-energy-problems. Accessed 21 Mar 2015
12. New York Times (2011) Solyndra Bankruptcy Reveals Dark Clouds in Solar Power Industry. http://www.nytimes.com/gwire/2011/09/06/06greenwire-solyndra-bankruptcy-reveals-dark-clouds-in-sol-45598.html?pagewanted=all. Accessed 21 Mar 2015
13. MIT Technology Review (2012) Can ARPA-E solve energy problems? http://www.technologyreview.com/news/427136/can-arpa-e-solve-energy-problems. Accessed 21 Mar 2015
14. Butler N (2015) How oil and gas majors are rethinking on climate change. http://blogs.ft.com/nick-butler/2015/04/06/oil-and-gas-majors-are-rethinking-on-climate-change
15. The Economist (2014) E.ON and E.OUT. http://www.economist.com/news/business/21635503-german-power-producer-breaking-itself-up-face-future-eon-and-eout. Accessed 13 Mar 2015
16. Vasagar J (2014) E.ON split driven by technology. http://www.ft.com/cms/s/0/e15f7f32-7a34-11e4-9b34-00144feabdc0.html#axzz3UdJ0xw51. Accessed 13 Mar 2015
17. Nicola S (2014) E.ON split to fortify German clean energy transformation. http://www.bloomberg.com/news/articles/2014-12-01/eon-split-to-fortify-german-green-energy-transformation. Accessed 14 Mar 2015
18. Zyga L (2010) Japanese firm wants to transform the Moon into a giant solar power plant. http://phys.org/news/2010-06-japanese-firm-moon-giant-solar.html. Accessed Feb 2015
19. Karajan, von H (2014) The Karajan register. http://karajan.org/jart/prj3/karajan/main.jart?rel=en. Accessed 10 Dec 2014

Part IV
Capital Markets, Time to Graduate from the Casino

To know values is to know the meaning of the market.
Charles Dow[1]

[1]Thornton ME (2011) 10 fantastic quotes about markets and money. http://www.businessinsider.com/10-great-quotes-about-the-market-2011-6?op=1&IR=T. Accessed 27 May 2015.

Bringing the Power of Financial Innovation to Bear

13

13.1 Matrix Reloaded

Next to kicking off the 2008–2012 global recession, the Global Financial Crisis of 2007–2008 wiped out as much as 45 % of global wealth. The public easily found whom it considered the culprits: greedy speculators and bankers. People came to view them as the proverbial "bad guys" who had invented new financial products that mispriced risk, only to cream off exorbitant profits [1].

To stave off the collapse of large financial institutions carrying systemic risk, the governments that could bailed out the key banks operating in their jurisdictions. In addition, governments took on massive amounts of sovereign debt to stimulate the economy.

They also tightened regulation to protect consumers. For example, the EU's Markets in Financial Instruments Directive 2004/39EC (MiFID) and Alternative Investment Fund Managers Directive (AIFMD) requires fund managers who want to sell their products in the European Union to register or even have representatives in all countries in which they want to sell their financial products [2].

Compared to the long supply chains in the real economy we looked at earlier—where manufacturing in emerging markets is often lightly regulated and voluntary standards have largely taken over—finance is more like the energy industry: highly regulated. This can have perverse side effects. For example, because of the stringent "corset" of regulatory constraints placed upon investment fund structures, more and more organize themselves via company holding structures. This is really a regression in terms of governance: responsibility of oversight and management risk are confounded, rather than separated.

Responding to the obvious regulatory deficiencies that had become apparent during the financial crisis, both global voluntary frameworks such as the Third Basel Accord and national governments tightened the screws [3]. At the bank level, fresh micro prudential regulation sought to raise the resilience of individual banking institutions to periods of stress. This by reducing bank leverage, increasing liquidity, and generally improving risk management, governance and transparency.

M. Martin, *Building the Impact Economy*,
DOI 10.1007/978-3-319-25604-7_13

Complementing these efforts, governments also introduced new macro prudential regulation to prevent system-wide risks from building up across the banking sector. Although banking crises are as old as the invention of banks, a further policy goal going forward was to better manage the procyclical amplification of such risks across the business cycle.

Regulation took off to such an extent that 83 % of respondents to the 2015 PricewaterhouseCoopers (PwC) financial services global CEO survey were "extremely" or "somewhat" concerned that growth could be threatened by over-regulation (though better regulatory coordination can also increase cross-border capital flows) [4]. At any rate, 69 % of asset management CEOs assumed that regulation would disrupt their industry over the next five years [4]. The concern: regulation will divert management attention from other strategic challenges.

Against this global context, three other trends advanced more or less quietly.

First, the digital disruption of capital markets powered on. Do-it-yourself and mobile banking as well as the digitization of transactions are the obvious stories. The digital disruption ahead will cut much deeper than this though. In the same global financial CEO survey, a majority of respondents stated that market entrants were disrupting their existing business models. They managed to better serve customer needs at distinct points of the financial services value chain by leveraging technology [5]. Anyone who can use technology to provide a better customer experience, at lower cost, and is unencumbered by legacy infrastructure and business models has the potential to become a formidable competitor.

The prospect of much more radical changes made its first appearance as well. Consider Bitcoin, a peer-to-peer (P2P) online payment system invented by a group of cypher punk developers in the middle of the financial crisis. They published an open-source cryptocurrency protocol under the pseudonym Satoshi Nakamoto. The viability and efficiency of modern payment systems is based on trusted, central third parties for secure processing of payments. However, the digital revolution now allows for new decentralized systems that could ultimately be cheaper and more secure. The leading technology worth exploring today is the Bitcoin and its blockchain technology.

In 2014, UBS' group chief information officer Oliver Bussmann argued that the blockchain could "not only change the way we do payments but it will change the whole trading and settlement topic." He viewed it as the single technology with the greatest potential to disrupt financial services overall, triggering a "massive" simplification of banking processes and cost structure [6]. The Bitcoin eliminates the need for an intermediary altogether: network nodes verify the transaction and record it in a public, distributed online ledger called blockchain, which uses its own unit of account, the Bitcoin [7]. The blockchain holds the prospect to revolutionize finance, as discussed below (Spotlight 12).

The redesign of processes and products and services following from digital technology enabled a second trend: the democratization of access to capital—via crowdfunding. In crowdfunding, some project or venture raises financial contributions online from the public. For the money contributed, the crowdfunder

(1) pre-purchases some product or service, thus enabling the launch of a new business concept by building a sales pipeline (so-called "rewards crowdfunding"). Alternatively, (2) the crowdfunder buys shares in an early stage company (so-called "equity crowdfunding"), or (3) makes a loan to a business or person online ("crowdlending") [8].

Other business models are litigation crowdfunding, where individuals back a plaintiff in a legal dispute against a stake in the claim funded, and the growing segment of charity crowdfunding where individuals donate to charitable projects online. For example, in its—admittedly, not entirely serious—overview "the top 10 legal disputes ripe for crowdfunding in February 2015," website Invest4Justice suggested to "Crowdfund a lawsuit to support gay marriage [...] If a lawsuit to support gay marriage is crowdfunded, then crowdfund a lawsuit to support gay divorce" [9].

The idea underlying crowdfunding has been around since before the Industrial Revolution: in the seventeenth century, authors financed book prints by asking members of the public to donate. In return, they offered a mention on the book's title page [10]. However, crowdfunding needed the Internet to take off in earnest. The nascent crowdfunding industry is well advanced in the US. Seeking to stimulate capital formation amongst small companies and startups, the JOBS Act of 2012 included a section on investment crowdfunding. Fresh avenues to mobilize capital are good, but concerns included that people would invest more than they could afford [11].

To address them, several principles were established: first, to limit the sale of securities over a one-year period (one million dollars per year); second to limit individual investors' exposure (the lesser of USD 100,000 or 10 % of annual income); third, mandatory disclosure of risks; and finally, the need to use a SEC-registered funding portal [12]. Regulatory implementation as per the JOBS Act Rulings by the U.S. Securities and Exchange Commission (SEC) remains complicated [13]. In 2015 the SEC rulings allowed non-accredited investors—the public at large—to participate in equity crowdfunding and investment in private startups and small businesses [14].

Other countries actively seeking to enable crowdfunding are Israel and the UK. With the Regulations of the Financial Conduct Authority (FCA), the UK has a comprehensive regulation, and caps investor exposure at ten percent of their "net investible assets" for so-called "inexperienced investors," based on a self-declaration.

In Israel, draft regulations were published by the Israeli Securities Authority in March 2014 [15]. In the need to balance between the needs of entrepreneurs and the needs of investors, "startup nation" Israel is generally taking a more restrictive approach. This goes a long way to protect the investor, allowing only hi-tech startups to pursue the route of crowdfunding, but can enough capital be mobilized? [15]. Moreover while Israelis excel in "hardcore tech," building chips, cyber technology and big data companies, the country's entrepreneurs seem to be weaker in slick consumer products that sell based on style [16]. Fascinating new platforms are coming up, and time will tell if they will eclipse the traditional much more local world of angel investing. For example, leveraging Israel's strong base of tech innovation and founded by a tech entrepreneur, platform OurCrowd selects startups and then invests its own capital, as well as opening its membership of 4000 accredited investors (who must be able to invest at least USD 10,000 in a deal) [17].

Crowdfunding platforms such as Kickstarter, Indiegogo, and Crowdfunder are currently the most widely known sites. In 2015, there were already more than 500 crowdfunding platforms globally. In the first half of 2014 alone, the CrowdDataCenter reported more than 4 million backers pledging over USD 357 million, and funds raised of over USD 322 million. At the time, global crowdfunding was reported to grow with around 500 new projects per day. 18,000–22,000 projects were open for funding at any given time [18]. Crowdfunding platforms are reported to have raised some USD 2.7 billion and successfully funded over a million campaigns in 2012, with an 81 % increase to USD 5.1 billion forecast for 2013 [19]. According to infoDev, the global crowdfunding market in developing countries could grow to USD 90–96 billion by 2025: more than 344 million people have incomes above USD 10,000 and at least three months of savings (or three months of savings in equity holdings) [20]. China holds a USD 50 billion crowdfunding potential [20]. The potential has been estimated to amount to about 1.8 times the size of the current global VC industry [21].

How fast the crowdfunding capital market ultimately grows depends also on regulation [22]. For example, the EU needs to figure out how to overcome national crowdfunding regulations, which are not harmonized. They impose burdens on European crowdfunding businesses, which operate cross-border, thereby restricting market entry. This at a time when the power of digital transcends national markets. Discrimination is not compatible with the guiding principle of establishing a European single market, and restricting crowdfunding as an alternative method of financing at the European level in the long-term is unlikely [23]. The EU has taken first steps toward formulating a legal framework for crowdfunding at EU level [24]. Progress may be uneven, but it is safe to assume that digital is ultimately unstoppable. Democratization of access to capital markets for the public at large will be a growing trend rather than a short-lived fashion.

Finally, a third idea also gained currency in the 2000s: that an investment could consciously achieve both a financial return *and* create some positive social or environmental externality. Again, in some ways, there was nothing new about this: for centuries, family business owners have kept creating value with a holistic mindset, considering both financial opportunity as well as social needs. In the process, they created important innovations such as health insurance schemes and pension funds for their workers and their families. However, the differentiating new idea that now gained ground was that such impacts could be systematically measured and monitored.

The reader will note that we are now entering the homestretch of the book's analysis of the key theaters of change to keep the global economy viable by making it sustainable. At the beginning of the twenty-first century, the world of finance has traveled quite a road from the Florentine merchants who invented double entry bookkeeping at the end of the thirteenth century. More than 600 years ago, these pioneers laid the foundations to price risk and allocate capital to investment opportunities in ways that were instrumental for the onset of the modern market economy and globalization. Today's pioneers seek to operationalize both new technology and our values so that investment decisions minimize adverse side effects, seize the potential of positive change, and price risk efficiently in all sorts of endeavors enabled by our globalized economy.

As we delve deeper into capital markets, one point is on order. In the demonization of bankers, what sometimes gets lost is that while a lot has changed since banking was an admired activity in Italy's wealthy trading cities in the Middle Ages and early Renaissance, pricing risk and allocating capital *continue* to be key activities to enable human progress. We are certainly no longer talking about financing ships to go to Indian ports and trade in spices. Notwithstanding, financial innovation is every bit as necessary as before.

Just consider the tremendous opportunity for innovation identified in the previous chapters. These new activities now need to be financed—when companies upgrade their supply chains or we invent and develop new energy technologies, such activities require investment before they can bear fruit. Rather than going away, as some wish simplistically, the need for modern finance will become more pronounced as the future unfolds. Given the magnitude of adjustment that lies ahead, next to technology, a key enabler is the transformation of the world of investments. Financial markets may indeed be out of control and resemble a big casino today, as some argue. Our job is to help bankers rebuild finance so they can fulfill their function as efficient capital allocators in ways that suit our needs in the twenty-first century.

Spotlight 12: Blockchain, Payment Revolution Next?

Is the Bitcoin a "cryptocurrency" without intrinsic value? [25] Not quite. In September 2014, the Bank of England publicly recognized the blockchain's potential, arguing that the technology could have "far-reaching" implications [26]. The potential lies in adapting blockchains to record and verify financial transactions, including clearing securities or making cross-border payments. How does this work? The Bitcoin blockchain provides a shared public ledger [25]. The ledger is open to all via a network providing a pretty much tamper-immune data structure. Once a transaction is confirmed, it is embedded in the Bitcoin blockchain (Fig. 13.1).

SHA-256 class system security cryptography then ensures the integrity of the blockchain applications: all transactions must be signed using a private key; Bitcoin miners then confirm transactions in the network within around 10 min [25]. Via a shared consensus system, several independent confirmations enable the transaction to go forward. The process uses random distribution, and this renders manipulation very difficult: even though it would be theoretically possible to compromise the network, the gigantesque size of the network and the corresponding resources required for an attack make this hardly feasible in practice [25]. Given the potential, the UK has started to establish itself as the global hub for Bitcoin trading. In 2015, HM Treasury announced it would begin regulating digital currencies by applying anti-money laundering ordinances to digital exchanges [27]. Further, the UK

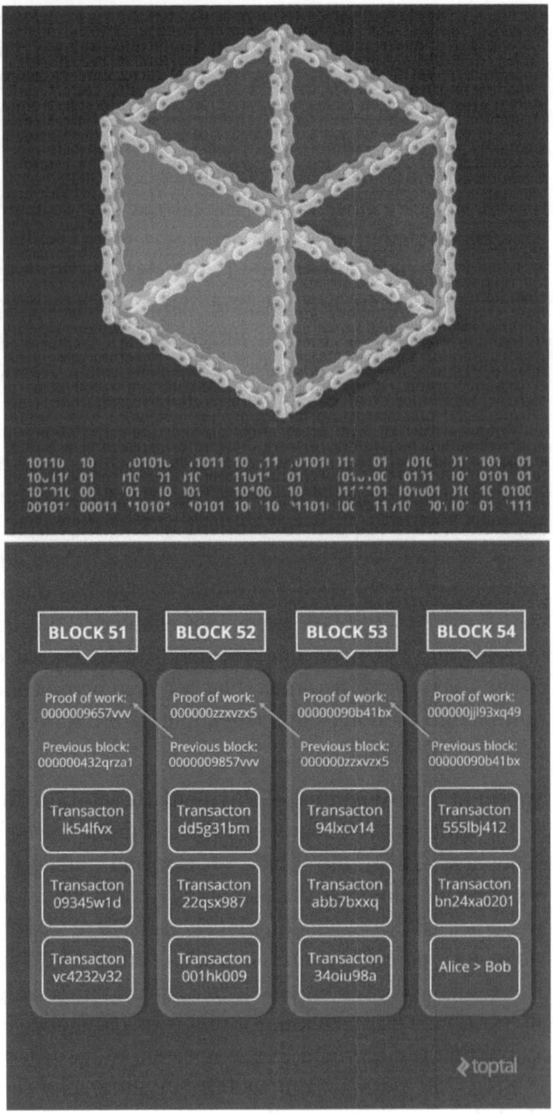

Fig. 13.1 Confirmation in the Blockchain (*source* Hajdarbegovic [25]) (Figure courtesy of Nermin Hajdarbegovic)

government committed GBP 10 million to finance research in digital currencies technology [28]. In the UK's fertile environment for Bitcoin, UBS went a step further in April 2015 and announced it would open a London-based technology lab to assess how blockchain technology could be used in financial services. It was the first bank to do so [29].

This is quite a departure after first being looked at as another Internet curiosity for geeks, to only then gain traction among groups including hackers, anarchists, libertarians, and darknet drug dealers—but entirely a result of the blockchain's enormous potential [25]. A cryptographic blockchain could be used to digitally sign sensitive information, and to thus decentralize trust. The Bitcoin blockchain has many applications including authentication, smart contracts and escrow services, or tokenization; this has triggered interest from mainstream information technology players such as Dell or Microsoft [25]. Other banks are exploring so-called "fin-tech" (financial technology) businesses as well. Barclays runs a fintech incubator. Most major banks participate in Accenture's fintech accelerator program. Lloyds Banking Group backs Startupbootcamp FinTech, and HSBC, Santander and BBVA launched corporate venture funds to make equity investments in fintech companies [29]. While we do not know whether regulation will ultimately enable or stifle its development, the Bitcoin is an excellent illustration of the disruption that digital has in store for financial services. We could be in for a wild ride, just like in the early years of the Internet [30]. While it is difficult to make predictions in this 'Wild West,' one thing is virtually certain: we will hear a lot more about the Bitcoin in years to come, not the least because the MIT Media Lab launched its Digital Currency Initiative in 2015, keen to join the race [31, 32].

13.2 Fresh Investment Content Wanted

Next to digitization, regulation, and values, another major theme in capital markets is the rise of emerging markets and especially Asia. For sure, one should not discount major financial hubs such as London and New York. They offer important secondary advantages to public markets, including the depth of legal expertise and the presence of major private equity funds. Yet by 2025, most large companies will consider a secondary listing in emerging markets, where the lion's share of global economic growth is taking place [33]. Less clear than such geographical trends though is where massive volumes of innovative investment content will come from. Awareness of the long-term environmental and social sustainability of the global economy is rising. This is not just changing the way people interact, but also how we allocate capital: investing according to environmental, social and governance criteria (ESG) is gaining ground. As LPs, institutional investors increasingly ask the major investment houses how they deal with sustainability topics.

For major institutional investors, the potential stranding of their assets is gradually becoming more and more important. Just like acting on the dangers of asbestos by banning its use in electric and building insulation disrupted these industries from the 1990s, removing fossil fuels from the energy mix has implications for the value of an energy company's oil or coal reserves. By the mid-2000s, 40 countries had completely banned asbestos, including most countries in the European Union [34].

There are numerous theaters of action, and infrastructure is one of them. In an advanced economy such as the US, the average American family spent one third more on transportation than on food in 2010, or USD 8600 a year [35]. Not only does this have important welfare effects—for those who are not in the top decile of income, transportation absorbs one out of every six dollars earned—but an inefficient or outdated infrastructure also has wider growth and sustainability implications [35]. Moreover, the former dichotomy between "advanced" economies and "emerging" markets is becoming increasingly blurred. Consider, for example, the country from where the ships departed that first incorporated Japan and Brazil in the global trading network five hundred years ago. Contemporary Portugal has a number of unique assets, including its privileged access to the 494 million consumers in the European Union and the 220 million consumers in the Portuguese-speaking countries around the world [36]. In 2013, its economy came out of its deepest recession in more than forty years, with ten quarters of economic contraction [37]. Youth unemployment shot up twenty percentage points in the crisis [38]. Portugal urgently needs to restore its long-term competitiveness and achieve inclusive growth [39]. Investment capital plays an important role in achieving this. Twenty years ago, there was no continuous motorway between Lisbon and Porto [39]. Since then, life expectancy has increased by ten years; infrastructure has made a significant leap forward, in many cases to world class [39]. The question is: how can the country build on its accomplishments during the past three decades, rather than dismantling them across the board to compete on cost? To create the urgently needed new jobs that were lost in the crisis, Portugal needs new sources of comparative advantage other than market access. Further reducing the level of public good provision on education and health—to name but two examples—moreover risks taking social cohesion beyond the breaking point.

Solutions that combine money with financial innovation are particularly promising. For example, financial instruments could be designed to embed incentives driving micro-level improvements that touch lives and boost competitiveness such as lowering the cost of grade repetition—which currently imposes an additional burden of more than ten percent on Portugal's total primary and secondary education expenditure. To stay in the field of education, financial engineering could also help bring down the cost of boosting Portugal's level of educational achievement from currently only 32 % of high school diplomas in the 25–64 years age group up to the OECD average of 74 % [40]. In the mainstream of the economy, Portugal similarly needs financial instruments to fund the upgrading and reactivation of the country's key industries. From 2007–2013, its fashion industry lost one third of its jobs [41]. Key in all cases is to link financial incentives to the continuous improvement of social outcomes, and the intelligent combination of different sources of capital, appropriately rewarding risk-taking.

What's needed are fresh solutions to finance the creation of opportunity for all. To deliver on capital formation, business opportunities and new jobs, social inclusion and effective operational implementation in Portugal and elsewhere, functioning capital markets considering such impacts and investable financial products are required. Integrating the possibilities the digital revolution offers with investing in ways that align values with value is hard work. None of this will happen spontaneously. Let us examine which contribution impact investing could make.

13.3 Turn Impact Investing into the Solution It Could Be

To bring the power of financial innovation to bear on the great challenges of our times we need to upgrade capital markets. In principle, the quest for holistic value opens the path to use capital much more efficiently to fund solutions to social and environmental challenges.

Which potential does then one much talked-about pathway, the market for so-called "impact investments" hold? Its proponents argue that investments made with the intention to generate measurable social and environmental impact alongside a financial return are a great tool to grow the economy and fund the provision of public goods [42]. If the renewables industry discussed in Part III is well described by the metaphor of field mice in need of growing up to become 800-pound gorillas, this is also true for impact investing. In fact, given current industry size, the genetic engineering required is even more pronounced.

The term "impact investing" was coined in 2007. The Global Impact Investing Network (GIIN), the interest group comprised of funders who identify with the concept and label of impact investing mentioned earlier, defines it as follows: "investments made into companies, organizations, and funds with the intention to generate measurable social and environmental impact alongside a financial return" [42]. The word "intention" in this characterization differentiates these investments from mainstream investments, and even from socially responsible investments (SRI). The latter screen for *avoidance* of social or environmental harm after the fact, emanating from an asset that is otherwise managed for a single bottom line: profit.[1] An impact investment can be made in both developed and emerging countries. It can either seek below-market or market-rate financial returns, depending on the investor's intention. This characteristic makes it different from a grant, which is simply a donation of funds with no expectation of financial returns. Program-related (PRI) and mission-related investments (MRI) are types of impact investments made by charitable foundations with different rules governing their use [43].

The potential returns of impact investments in emerging markets sound particularly compelling. A 2010 study by J.P. Morgan, based on a survey of impact investors, found that the expected returns of many existing impact investments in emerging markets fell largely in the 8–11.9 % bracket for debt investments, and in the 20–24.9 % bracket for equity. This compared to developed market return expectations of 5–7.9 % and 15–19.9 % in debt and equity respectively [44].

Such return projections sound exciting when compared to financial return expectations in the low-growth environment of the European Union in mainstream financial markets immediately following the financial crisis. However, as an investment style, impact investing is too recent and the classification of investments as impact investments (or not) too nascent to have any definite views about return expectations. Notwithstanding, it is important to retain that impact investing is not

[1] This book uses "social impact investment" as coterminous with "impact investment," unless otherwise specified.

simply an emerging markets phenomenon. In the UK, the RBS Social Enterprise 100 index entered its sixth year in 2015; it aims to track the success of social businesses across the UK, and examines both their growth and the impact that they make [45]. The SE100 Index not only showed high average growth, but the growth in revenue by index entrants had been impressive when compared with both the Financial Times Stock Exchange 100 Index (FTSE100) and the SME sector [46].

When measured by the cumulative investments outstanding of the members of the GIIN, this type of investments had grown into a USD 60 billion market by 2014, with commitments of USD 12.2 billion in 2015 [47]. This is remarkable progress since the inception of the foundational term "impact investment" in 2007, and with a mere USD 4.3 billion worth of impact investments made in 2011 [47]. Moreover, not every impact investor is a member of the GIIN.

In the big scheme of things, impact investing is still a minuscule phenomenon in financial markets. However, experts estimate a potential to advance the overall pool of assets invested for impact to USD 400–1000 billion by 2020 [48]. This would represent exponential growth. The worrying insight though is that even one trillion dollars would still amount to "peanuts." Just relate one trillion dollars to today's total global capital stock—as measured by bonds and stocks—of USD 212 trillion, with stocks at USD 54 trillion [49]. Put in a different perspective, the tail expected to wag the dog is even smaller. In 2010, USD 600 trillion in all financial assets represented claims against a nominal global GDP of roughly USD 63 trillion in the same year; by 2020, these assets are forecast to advance to roughly USD 900 trillion in financial assets, representing claims against about USD 90 trillion/year of global GDP [50].

To enable success in terms of fresh investment content means liberating impact investing from today's niche existence. To take the major step forward that is now needed it is indispensable to engineer a sound investment ecosystem composed of quality investors, investees and intermediaries, and a supportive regulatory agenda, which are discussed in the chapters that follow. Further, impact investing needs to take advantage of the trends disrupting finance such as digitization or crowd-funding. In many jurisdictions, regulation is on a trend to make that task harder rather than easier. Delivering on the potential will require new investment ideas as well as reaching outside the current networks of self-identified impact investors, given their current pace of building scale. To matter, impact investments need to become a part of standard asset allocations in investor portfolios. This raises two *how-to* questions [51]:

First, given the still relatively minor "stock" of impact investments around the world, how can this be achieved in practice?

Second, like batteries in e-mobility, what are the critical constraints standing in the way of scale?

From the pace of advancement of innovations in other fields, we know that the impact may be—and to matter has to be—much larger than suggested by current growth projections. Time to give it a shot and create more holistic financial propositions that integrate extra-financial outcomes as well, regardless whether we call this approach "impact investing" or in some other way. ESG has already made a relevant contribution. Let us keep in mind the compliance logic governing brands' interactions with garment manufacturers in emerging markets discussed earlier. It proved insufficient for wide-ranging supply chain upgrading. In finance, a real solution will similarly have to achieve more than policing existing assets via ESG reporting and positive and negative screening of public stocks—with information about extra-financial performance largely confined to listed firms in advanced economies anyway. Thus far, ESG investing has often ended up producing institutional investment portfolios whose positions look very similar to traditional unscreened portfolios, except their exclusion of assets in domains such as arms or tobacco. We need to do better than that.

For several reasons, we can expect fresh solutions to come on stream in the near future. First, there is demand out there—because there is real work to do. Despite the benefits reaped from globalization and the technology revolution, many of humanity's same old social problems persist, and environmental problems are getting worse. Countries have spent trillions of dollars combatting these challenges, yet evidence suggests that this strategy is not working all that well. Second, while the pace of human population growth by 2050 is forecast to fall to by more than half compared to its peak in 1989, we can nevertheless expect to get to a global population of 9 billion by mid-century. This is not to mention the implications of an increasingly aging population, owing to wider access to more and better information about healthy living along with advances in medicine [52].

With fewer people working and more people living well into their eighties, nineties, and longer, our social systems, and the problems they address, will become much harder to fund. A 2012 study by leading global forecasting and quantitative analysis firm Oxford Economics projected a sizeable expenditure gap in 2025 to pay for delivering public services at existing levels in 2025. For Brazil, the gap amounted to USD 70 billion (1.9 %); for Germany, USD 80 billion (2.0 %); for India, USD 70 billion (1.5 %); for the US as much as USD 940 billion (4.4 %), and for the UK even USD 170 billion or 5.4 % of GDP [53].

To finance this gap and other challenges, new pathways and solutions are needed. Fortunately, there are many examples of financial innovation to be inspired by. For starters, similarly to the issues of global supply chains and the energy trilemma considered earlier, a shift in mindset is useful. Take microfinance—or financial systems at the bottom of the economic pyramid that broaden access to capital for individuals with low income. Over the past 20 years, microfinance has transformed how we view the poor: rather than non-bankable recipients of assistance, they are productive micro-entrepreneurs in the informal sector, capable of building wealth and taking part in the mainstream of the economy. As an asset class with USD 81.5 billion in outstanding loans (as of 2014), microfinance has influenced international development and financial regulation in emerging economies [54]. The approach has driven much more significant progress than the mere size of outstanding loans would suggest [55]. A decade ago, microfinance banks even started to penetrate the mainstream via IPO on stock exchanges, as in the case of Bank Rakyat Indonesia, which first listed on the Jakarta Stock Exchange in 2003, and Banco Compartamos in Mexico, which IPOed in 2007 [56].

Given major adjustment challenges ahead in domains such as education, healthcare, unemployment and the environment, investing with an impact mindset can potentially play a similarly catalytic role. This on an even more comprehensive scale. We need to remain creative and innovate though. The Bitcoin asks us to experiment with decentralized secure payment systems. This could disintermediate banks on a massive scale. Who knows whether our children's children will still have bank accounts. Analogously, we need to grasp that marrying big data and financial engineering is now enabling us to measure extra-financial outcomes and construct investment propositions that can achieve both financial returns and social impact. The opportunity is to leverage innovative entrepreneurs, trillions of public and private money, and financial innovation to benefit society. To make the most of turning our most enduring social challenges into opportunities, impact investing needs to position itself to benefit from forces such as digitization, shifting regulation, and the rise of emerging markets.

To matter, we need to be ambitious. This means taking impact investing to a critical mass where it can make a noticeable contribution in the development trajectories of countries. The cheapest and fastest way to build such a capital market is a mix between private initiative and competition and collaborative, systemic action. Examples of market building efforts around the world, including the lessons from the pioneering UK experience, are instructive. The argument put forth in the following chapters emanates from well over a decade of work in the merging of the investment and the impact field, including taking part in the original meeting in Bellagio, Italy, where the term "impact investing" was first collectively coined. Research activities for Part IV included a screening of almost 200 impact investment-related reports in the major languages of the G8 and a dedicated online survey distributed to over 250 market players from around the world in 2013. Commissioned by the UK Cabinet Office, an earlier partial version of Part IV was circulated to participants of the first-ever G8 conference on social investment held in 2013 under the UK presidency of the G8 in that year, as a companion to my in-depth report "Making Impact Investible" published by the investment and strategy firm Impact Economy [57, 58]. To appreciate a possible role of capital markets in a post-casino paradigm, let us now place impact investing in the broader context of the four megatrends to which we need to adjust (introduced in Part I).

References

1. Megan D, Walden S (2009) 45 percent of world's wealth destroyed: Blackstone CEO. In: Reuters. http://www.reuters.com/article/2009/03/10/us-blackstone-idUSTRE52966Z20090310. Accessed 6 Jun 2015
2. European Commission (2011) Alternative Investment Fund Managers Directive 2011/61/EU. In: Official Journal of the European Union, L 174/1. http://ec.europa.eu/yqo/index.cfm?fuseaction=legislation.showlexld=9. Accessed 4 Sept 2014
3. Bank for International Settlements (2015) International regulatory framework for banks (Basel III). http://www.bis.org/bcbs/basel3.htm. Accessed 9 Jun 2015
4. PwC (2015) Asset management 18th annual CEO survey 2015. http://www.pwc.com/gx/en/ceo-survey/2015/industry/asset-management.jhtml. Accessed 30 May 2015

5. PwC (2015) Financial services regulation, updates on AMLD, CEO survey and financial models. http://www.pwc.com/gx/en/financial-services/european-financial-regulation-updates/february-16-2015.jhtml. Accessed 30 May 2015
6. Irrera A (2014) Innovation spaces where CIO lives the dream. In: Financial News. http://www.efinancialnews.com/story/2014-10-28/fintech-news-2-oliver-bussman-ubs?ea9c8a2de0ee111-045601ab04d673622ea9c8a2de0ee111045601ab04d673622. Accessed 15 Jul 2015
7. Nakamoto S (2015) Bitcoin: a peer-to-peer electronic cash system. http://bitcoin.org/bitcoin.pdf. Accessed 15 May 2015
8. CrowdDataCenter (2014) Mapping the state of the crowdfunding nation. http://thecrowdfundingcentre.com/downloads/eFunding_-_The_State_of_The_Crowdfunding_Nation_-_Q2_2014_HEADLINE_EDITION.pdf. Accessed 15 Jul 2015
9. Invest4Justice (2015) The top 10 legal disputes ripe for crowdfunding in February 2015. https://www.invest4justice.com/the-top-10-legal-disputes-ripe-for-crowdfunding. Accessed 15 Jul 2015
10. Taylor B (2013) The basics of crowdfunding: how crowdfunding works. Factory. http://blog.ideascale.com/tag/history-of-crowdfunding. Accessed 15 Jul 2015
11. Kassan J (2013) The JOBS Act leaves crowdfunding investors unprotected: the SEC is working with a flawed law. http://venturebeat.com/2013/10/31/the-jobs-act-leaves-crowdfunding-investors-unprotected-the-sec-is-working-with-a-flawed-law. Accessed 15 Jul 2015
12. Shenhav A, Hoffman G (2014) Crowdfunding: current status and proposed changes. http://www.shenhavlaw.com/wp-content/uploads/2014/06/Crowd-Funding-Presentation.pdf. Accessed 15 Jul 2015
13. U.S. Securities and Exchange Commission, Division of Trading and Markets (2012) Jumpstart Our Business Startups Act: frequently asked questions about crowdfunding intermediaries. http://www.sec.gov/divisions/marketreg/tmjobsact-crowdfundingintermediariesfaq.htm. Accessed 15 Jul 2015
14. Barnett C (2015) SEC democratizes equity crowdfunding with JOBS Act Title IV. In: Forbes. http://www.forbes.com/sites/chancebarnett/2015/03/26/infographic-sec-democratizes-equity-crowdfunding-with-jobs-act-title-iv. Accessed 18 Jul 2015
15. Shenhav A, Hoffman G (2014) Crowdfunding—current status and proposed changes. http://www.shenhavlaw.com/wp-content/uploads/2014/06/Crowd-Funding-Presentation.pdf. Accessed 15 Jul 2015
16. Crowdfunding Israel (2013) Why isn't the startup nation the crowdfunding nation? http://www.crowdfundisrael.com. Accessed 15 Jul 2015
17. Fannin, R (2014) Crowdfunding could top angels If Israel's OurCrowd keeps pace. In: Forbes. http://www.forbes.com/sites/rebeccafannin/2014/07/15/crowdfunding-could-top-angels-if-israels-ourcrowd-keeps-pace. Accessed 15 Jul 2015
18. CrowdDataCenter (2014) Mapping the state of the crowdfunding nation. http://thecrowdfundingcentre.com/downloads/eFunding_-_The_State_of_The_Crowdfunding_Nation_-_Q2_2014_HEADLINE_EDITION.pdf. Accessed 18 Jul 2015
19. Massolution (2013) CF crowdfunding industry reports. http://research.crowdsourcing.org/2013cf-crowdfunding-industry-report. Accessed 15 Jul 2015
20. InfoDev (2013) Crowdfunding's potential for the developing world. http://www.infodev.org/infodev-files/infodev_crowdfunding_study_0.pdf. Accessed 15 Jul 2015
21. Noyes K (2014) Why investors are pouring millions into crowdfunding. http://fortune.com/2014/04/17/why-investors-are-pouring-millions-into-crowdfunding. Accessed 15 Jul 2015
22. European Crowdfunding Network (2014) Review of crowdfunding regulation: interpretations of existing legislation concerning crowdfunding in Europe, North America and Israel. http://www.osborneclarke.com/media/filer_public/61/fe/61fe40e6-5790-41d0-b181-bb2065bca9a8/ecn_-_review_of_crowdfunding_regulation_2014_3.pdf. Accessed 15 Jul 2015
23. European Citizen Action Service (2014) Crowdfunding—need for EU regulation? http://www.ecas.org/crowdfunding-need-eu-regulation. Accessed 15 Jul 2015

24. European Crowdfunding Network (2013) Review of crowdfunding regulation 2013. http://www.eurocrowd.org/2013/10/review-crowdfunding-regulation-2013. Accessed 15 Jul 2015
25. Hajdarbegovic N (2015) Blockchain technology explained: powering bitcoin. http://www.smartdatacollective.com/nermin-hajdarbegovic/301341/blockchain-technology-explained-powering-Bitcoin. Accessed 15 Jul 2015
26. Robleh A et al (2014) Innovations in payment technologies and the emergence of digital currencies. http://www.bankofengland.co.uk/publications/Documents/quarterlybulletin/2014/qb14q3digitalcurrenciesBitcoin1.pdf. Accessed 15 Jul 2015
27. Webber D (2015) The transatlantic battle to regulate the Bitcoin market. http://www.fxcompared.com/blog/the-transatlantic-battle-to-regulate-the-Bitcoin-market. Accessed 18 Jul 2015
28. HM Treasury (2015) Digital currencies: response to the call for information. https://www.gov.uk/government/uploads/system/uploads/attachment_data/file/414040/digital_currencies_response_to_call_for_information_final_changes.pdf. Accessed 18 Jul 2015
29. Irrera A (2015) UBS to open blockchain research lab in London. In: Wall Street Journal. http://blogs.wsj.com/digits/2015/04/02/ubs-to-open-blockchain-research-lab-in-london. Accessed 18 Jul 2015
30. Joi I (2015) Why Bitcoin is and isn't like the Internet. http://joi.ito.com/weblog/2015/01/23/why-Bitcoin-is-.html. Accessed 18 Jul 2015
31. Gardiner M (2015) Launching a digital currency initiative. https://medium.com/mit-media-lab-digital-currency-initiative/launching-a-digital-currency-initiative-238fc678aba2. Accessed 15 Jul 2015
32. Joi I (2015) Announcing the MIT Media Lab digital currency initiative. http://joi.ito.com/weblog/2015/04/15/announcing-mit-dci.html. Accessed 18 Jul 2015
33. PwC (2011) Capital markets in 2025: the future of equity capital markets. https://www.pwc.com/en_US/us/transaction-services/publications/assets/capital-markets-2025.pdf. Accessed 15 May 2015
34. Tim Povtak (n.d.) The history of the ban on asbestos. http://www.asbestos.com/legislation/ban.php. Accessed 5 Jun 2015
35. U.S. Department of Treasury (2010) An economic analysis of infrastructure investment: a report prepared by the Department of Treasury with the Council of Economic Advisors. White House. https://www.whitehouse.gov/sites/default/files/infrastructure_investment_report.pdf. Accessed 5 Jun 2015
36. Aicep Portugal Global (2013) Investing in Portugal: business support services sector. http://www.portugaloutsourcing.pt/Docs/Conf8/Presentation_Innovation_Investissement_Portugal.pdf. Accessed 15 May 2015
37. Wise P (2013) Portugal emerges from its deepest recession in more than 40 years. In: Financial Times. http://www.ft.com/intl/cms/s/0/58975e76-04d9-11e3-9e71-00144feab7de.html#axzz3ck-ntoClk. Accessed 15 May 2015
38. The Economist (2011) The jobless young: left behind. http://www.economist.com/node/21528614. Accessed 15 May 2015
39. Martin M (2014) Making impact investible: preface to the Portuguese edition. http://www.impacteconomy.com/papers/IE_PREFACE_MII_PT.pdf. Accessed 15 May 2015
40. OECD (2012) Education at a glance 2012 highlights. http://www.oecd.org/edu/highlights.pdf. Accessed 15 May 2015
41. European Skills Council (2014) European sector skills council textile clothing leather footwear report 2014. http://europeanskillscouncil.t-c-l.eu/pdoc/22-eng/2014_report_F.pdf. Accessed 3 May 2015
42. Global Impact Investing Network–GIIN (2014). The GIIN register. http://www.thegiin.org/cgi-bin/iowa/resources/about/index.html. Accessed 11 Jul 2014
43. Levitt DA (2011) Investing in the future: mission-related and program-related investments for private foundations. http://www.adlercolvin.com/pdf/PTXL1105_Levitt.pdf. Accessed 1 May 2013

44. Arosio M (2011) Impact investing in emerging markets. http://www.sustainalytics.com/sites/default/files/impact_investing_in_emerging_markets-issues_for_responsible_investors.pdf. Accessed 1 May 2013
45. RBS SE100 (2015) Five years of analysis. https://se100.net. Accessed 1 May 2015
46. Social Enterprise (2011) The RBS SE100 data report 2011: charting the growth of the UK's top social businesses. Social enterprise and RBS community banking. http://www.socialenterpriselive.com/sites/default/files/files/supplements/SE100_DataReport_Web1.pdf. Accessed 1 May 2013
47. Saltuk Y, El Idrissi A (2015) Global social finance. spotlight on the market: the impact investor survey. J.P. Morgan and the Global Impact Investing Network. http://www.impactinvestingaustralia.com/countriesandregions/uk/j-p-morgan-giin-5th-annual-impact-investor-survey-launched. Accessed 1 May 2015
48. Global Impact Investing Network (2013) Perspectives on progress: the impact investor survey. http://www.thegiin.org/cgi-bin/iowa/resources/research/489.html. Accessed 1 May 2015
49. QVM Group (2012) World capital markets: size of global stock and bond markets. http://qvmgroup.com/invest/2012/04/02/world-capital-markets-size-of-global-stock-and-bond-markets. Accessed 28 May 2015
50. Bain & Company (2012) A world awash in money: capital trends through 2020. http://www.bain.com/Images/BAIN_REPORT_A_world_awash_in_money.pdf. Accessed 20 May 2013
51. O'Donohoe N et al (2010) Impact investments: an emerging asset class. http://www.rockefellerfoundation.org/uploads/files/2b053b2b-8feb-46ea-adbd-f89068d59785-impact.pdf. Accessed 1 May 2013
52. Wikipedia (2013) Population growth. http://en.wikipedia.org/wiki/Population_growth. Accessed 1 May 2013
53. Accenture (2012) Delivering public service for the future: navigating the shifts. http://www.youtheconomicopportunities.org/sites/default/files/uploads/resource/Accenture-Delivering-Public-Service-for-the-Future-112812.pdf. Accessed 1 May 2013
54. Jakubowicz J (2015) The future of microfinance: towards a new deal? Microfinance barometer 2015. http://www.european-microfinance.org/docs/news/BMF-2014-EN_web.pdf. Accessed 15 May 2015
55. Rhyne E (2010) The state of global microfinance: how public and private funds can effectively promote financial inclusion for all. Paper presented before the Subcommittee on International Monetary Policy and Trade, Committee on Financial Services, U.S. House of Representatives, 27 Jan 2010
56. Global Envision (2007) The history of microfinance. http://www.globalenvision.org/library/4/1051. Accessed 13 May 2015
57. Martin M (2013) Status of the social impact investing market: a primer. UK Cabinet Office. http://www.impacteconomy.com/de/primer1.php. Accessed 13 May 2015
58. Martin M (2014) Impact Economy launches the Italian edition of making impact investible. Impact Economy. http://www.impacteconomy.com/press/PR_MII_EN.pdf. Accessed 13 May 2015

Financing Our Adjustment to Megatrends

14

14.1 Finance Progress at the "Base of the Pyramid"

The term "Base of the Pyramid" (BoP; often used interchangeably with "Bottom of the Pyramid") was first coined in the 1930s. It has since evolved to describe the 4 billion people from mostly developing countries living on less than USD 2 per day. Since 1960, the World Bank (WB), donor nations, aid agencies, and others have expended well over four trillion dollars to eradicate the persistent problems of the BoP in areas such as health, finance, and housing [1]. Nevertheless, they have so far met limited success [2]. Official Development Assistance (ODA), a USD 135.1 billion industry in 2013, has come under criticism and shifted its focus toward greater aid effectiveness in the 2000s [3–6].

As a result, the door is now open for private enterprises to complement the work under way and bring innovative, market-based solutions to bear on these challenges, while simultaneously unlocking the approximately five trillion dollars in latent BoP demand. In so doing, they can stimulate their own domestic economies [7]. For example, sustainability-related global business opportunities in natural resources (including energy, forestry, agriculture and food, water and metals), health and education could build up steadily to around USD 3–10 trillion annually in 2050 at constant 2008 prices, or around 1.5–4.5 % of world GDP at that time [8]. Moreover, private enterprise can often help the poor "graduate" to the middle class by providing products at an affordable price. History suggests the middle class to be key in terms of providing the input for growth in an economy, such as new ideas and human and physical capital. In his book, *The Fortune at the Bottom of the Pyramid*, C.K. Prahalad describes how the poor, who live in high-cost ecosystems often run by slumlords and the like, pay a premium for everything from rice to credit. This so-called "poverty penalty" can be as high as 5–25 times what the rich pay and is a massive disintermediation opportunity [2].

Part of the funding for value creation in the BoP will come from traditional capital markets and established forms of public-private partnerships. An example of a company that is positioning itself to take advantage of this opportunity is the

M. Martin, *Building the Impact Economy*,
DOI 10.1007/978-3-319-25604-7_14

French electricity provider, Electricité de France (EDF). In 2002, it created a public-private partnership to profitably bring renewable energy to the nearly 10 % of Moroccan citizens who, because of prohibitive costs, were off the grid. By virtue of its presence in this market, EDF has gained important insight into developing countries' market dynamics. This will pay dividends as the company strives to bring solutions to challenges, energy or otherwise, faced by the BoP [9]. The growth in telecom in recent years discussed earlier is another example of an industry benefitting from developing countries' demand [10].

In many cases, companies who can provide desirable products to BoP populations can simply drive out inefficiencies and create profits. Some markets, however, are harder to serve and need layered investments to get off the ground. Healthcare provides an illustrative example. 39 million people globally are blind and approximately 246 million people have serious visual impairments [11]. 90 percent live in the developing world and 1–2 million people lose their sight each year [12]. Cataracts are the leading cause of preventable blindness and an estimated 75 % of these incidences are treatable or preventable [12]. There are over 120 eye care programs around the world that profitably offer quality services to middle class clients alongside free operations for the poor [13]. Properly run and scaled, such programs generate sufficient cash flow to support debt financing at manageable levels of risk and can access capital markets on quasi-commercial terms [14]. Impact-minded investments are instrumental in providing funding that allows for market building. For example, Deutsche Bank closed the Eye Fund in 2010, a USD 14.5 million fund launched in partnership with non-profit organizations, which finances expanded eye care services to the poor in developing countries [15]. Commercial investors receive a market rate of return. US foundations making program-related investments receive lower returns, but also achieve their mission by mobilizing non-grant capital for health outcomes. More recently, an innovative private equity fund designed to support the clinical development of new drugs and vaccines for diseases such as tuberculosis, malaria, rotavirus and cholera similarly used a tiered capital structure and reached a USD 108 million final closing [16]. Recipient of the 2014 Financial Times/IFC business award, the Global Health Investment Fund used a partial guarantee provided by the Bill and Melinda Gates Foundation to contain investors' potential loss at minus 40 %, even in the event of a total write down of the fund [17]. Such solutions can serve as a blueprint for other development challenges that can be addressed through market mechanisms.

In grasping the full potential of impact investment, it is important to remember that the financial services industry has the expertise to engineer a positive financial return even when the underlying asset loses money [18]. Innovative approaches to financing at the BoP create new opportunities to leverage scarce ODA and philanthropic resources. By funding first-loss tranches or issuing guarantees, they can serve as a catalyst for bringing additional private capital to the table in pursuit of development outcomes, alongside the massive commercial investments in the BoP already under way.

14.2 Invest in Total Resource Productivity

The annual percentage reduction in carbon emissions currently required to limit global warming to 2 °C is 4.8 percent per year. This figure is in excess of what has been proven historically achievable [19]. The World Economic Forum and Bloomberg New Energy Finance estimate the needed reduction will require low-carbon funding of USD 500 billion per year by 2020 [20]. The investment gap is substantial: actual investment in 2014 was estimated at USD 270 billion [21].

Whether or not the rate of global warming can be slowed, the needs of a global population estimated to reach 9 billion by 2050 will likely push natural resource limits, demanding for investments in new infrastructure, as well as much higher resource efficiency. The challenge is daunting. But, by adopting more of a long-term view, investments in energy efficiency can produce cost savings and promote job growth. For example, a study by McKinsey proposed 60 measures that Russia, with its diverse energy resources, can implement now while maintaining energy consumption and greenhouse gas emissions at their current levels [22]. The program would require EUR 150 billion in investments over the next twenty years, but would save up to EUR 345 billion over the same period [23]. A 2006 study argued that in Europe, one million jobs could be created from a 20 % cut in energy consumption, then the equivalent of EUR 60 billion annually [24]. This type of job stimulant could be a great opportunity for young people to shape the energy future of Europe. In 2014, unemployment among those aged 15–24 was 10.5 % in France and 22.7 % in Spain [25]. This figure includes those looking for work, with one estimate of those "not in employment, education or training" at 14 million. In 2014, both groups together cost member states an estimated EUR 153 billion a year in welfare benefits and lost production—1.2 % of the EU's gross domestic product [26].

While consciously investing for both financial and extra-financial returns simultaneously is about private initiative and risk-taking, governments set framework conditions and can crowd in or crowd out capital. Approximately half of the 20 % reduction in energy consumption could be achieved with a fuller embrace of European Community Directives already in place [27]. Some of the impediments so far concern a lack of information and training on the latest technologies. This can ultimately lead to risk-averse investors endorsing outdated technology even when it might not be the most efficient. As discussed in Part III, policies promoting transparency, education, and industry cohesion, all supported by broader financial incentives, could begin to reduce Europe's steep reliance on oil and gas imports, estimated to increase 0.6 percent each year, by 2030, based on current trends [28].

Ever since the signing of the Kyoto Protocol in 1997, substantial work has been under way in green finance. Even so, making the connection with impact investing and engaging corporations, investors, non-profit organizations, and governments in new ways could go a long way towards closing the investment gap. For example, R20 Regions of Climate Action—a non-profit organization founded in 2010 by former Governor of California Arnold Schwarzenegger—has formed a strategic partnership with the Asian Development Bank (ADB) to develop a project pipeline to be financed through various ADB facilities. R20 member regions, national governments and R20 partners such as the International Chamber of Commerce mitigate some of the risks associated with the potential projects. They also identify additional investors that would either co-invest with ADB or create their own syndicate for supporting low-carbon projects [29].

One of the areas where green finance can pay off fast is street lighting. Only about 1 % of electricity is used to light streets in the US (and about 1.3 % in the EU), but this ticket item can nevertheless command upwards of 60 % of municipal electricity spending in some areas [30]. Each year this accounts for 159 TWh of electricity globally, more than the annual output of three-dozen 500 MW power plants [31]. Advanced LED lights can cut energy usage, maintenance costs and reduce emissions, all while delivering better light quality for improved visibility and community or workplace safety. The costs of the LED lights can often be recovered from the savings in avoided electricity costs in 36–72 months [32]. Unfortunately, many other opportunities have much longer payback times.

Next to clear and stable regulatory parameters, investments in the green economy need mechanisms that bring transparency to the market by drawing together players that can aggregate demand on platforms where the risk can be defined, next to clear benchmarks of success to allow further injections of capital. Just as in the case of Eye Fund mentioned earlier, layered investment structures and public-private-non-profit partnerships are often a key to mobilizing new capital and achieving investment periods that are viable for investors [33].

14.3 Finance the Modernization of the Welfare State

Private investors are increasingly funding the provision of public goods. In some fields, this is already standard practice, such as the case of toll roads or airports. The concept can be extended to cover countries' key social challenges as well. Given the magnitude of the adjustment ahead, private capital is needed. The ratio of workers contributing to the welfare system to dependents—those benefiting from said welfare system—peaked for all advanced economies (North America, Europe, and East Asia) between 2000 and 2010 [34]. For these countries, the welfare state is no longer sustainable without major reforms. The issue of aging is particularly illustrative. Life expectancy has increased worldwide by 21 years since 1950. In addition, most developed countries are far below the so-called 2.1 replacement rate needed to maintain a stable population from one generation to the next. The following figures estimate total public benefits to the elderly as a share of GDP in percent in 2007, and what they are expected to be in 2040: Canada 8.3/14.7; France 16.6/23.5; Germany 15.8/21.7; Italy 18.0/24.7; Japan 14.1/18.4; Russia 5.8/10.2; UK 12.1/18.2; and US 8.9/16.3 [35]. Other challenges, such as poverty and recidivism, also prove persistent.

Fortunately, new financing mechanisms have begun to emerge to address these challenges. The UK, like many countries, struggles with the issue of recidivism. People who have already been through the criminal justice system commit about half of all crime. Reoffending costs the UK taxpayer an estimated to be GBP 9.5 to GBP 13 billion per year [36]. Piloted in 2010 and based on work on contingent financial return models since 2005, so-called SIBs shift risk away from government and onto other stakeholders. SIBs finance the service delivery upfront; they are only repaid if reduced reoffending rates occur [37]. The public sector benefits: it pays out to investors only for positive social outcomes, and only up to a fixed share of the cost savings achieved. Investors bear the risk that the funded measures do not lead to better social outcomes. Given their impact goals, investors also benefit since they can achieve a blended return composed of a social and a financial return. These social and financial return requirements are consistent in that the magnitude of the social return actually generated determines the financial return. If no positive outcome is achieved, the investor may lose his entire capital. Therefore, the SIB is, in actuality, a structured product where the underlying investment is spent on social services, and principal and return are paid out on the contingent achievement of a social objective. Social services benefit as well since the proceeds of the "bond issue" allow for paying the providers of social services upon delivery. They are thus encouraged to focus on results as opposed to just the absolute cost of their services. SIBs are useful under four conditions. These are: (1) the social problem is measurable and causes the public high costs; (2) the target group is clearly identifiable; (3) the proposed measures are known and tested; and (4) they cost considerably less than the cost savings they achieve. Given their direct involvement with public value creation, SIBs require favorable regulation and likely direct government incentives [38]. Setting up dedicated task forces and vehicles to pioneer the solution can also be instrumental. This was the case in the UK with the creation of Social Finance, a nonprofit organization dedicated to mobilizing capital for social progress. The developer of the first SIBs, it was initially financed by a group of private philanthropists [39].

SIBs have serious potential to make inroads in another major challenge: unemployment. It is well known that some retraining programs are better than others, but reallocating assets accordingly is often difficult. For example, take the German case of the "Joblinge" program [40]. Joblinge is a joint initiative by the Eberhard von Kuenheim Foundation and The Boston Consulting Group; it is a successful and cost-effective program for addressing youth unemployment, yielding a reinsertion rate of 70 % [41]. Even though Joblinge is much more cost effective than competitor programs offered by other providers and yields a higher reinsertion rate of unemployed youth, the program only covers a fraction of the 7.2 % unemployed youth that were awaiting labor market reinsertion in Germany in 2014 [42]. Instead, the bulk of the target group goes through less efficient programs. They cost more money and yield less impact. Given that the pilot Joblinge worked, we proposed in 2011 that further funding and replication could be addressed through a SIB. Based on reinsertion success and the resulting public budget savings and tax revenue instead of a transfer payment, the public sector could in turn financially reward private investors, provided the majority of the savings stay with the public sector. In 2013, the Joblinge SIB became a reality, when the Kuenheim Foundation launched the solution in collaboration with BHF-BANK Foundation, BonVenture, and BMW Foundation Herbert Quandt [43].

By the mid-2010s, SIBs came to be considered a flagship of new impact investment instruments aimed at keeping the welfare state fundable in the twenty-first century. To date, there is not yet enough evidence about the efficacy of SIBs to make any definite conclusions about them. Notwithstanding, they hold great promise to harness both public and private resources, and are part of a broader trend. Contingent financial instruments that produce a financial outcome dependent on achievement of a social objective are gaining traction. They can be expected to have significant implications for a broad range of sectors—from moderate-income housing, to health care, water and sanitation, and rural development. There are also attempts to apply the contingent return idea to social issues in developing countries. This has resulted in a series of so-called "Development Impact Bond" pilots or "DIBs" [44]. Pilot applications include funding measures to reduce sleeping sickness in Uganda, antiretroviral treatments to prevent HIV and TB in Swaziland, as well as access to education in Pakistan and Uganda [44]. Often subsumed under the heading "innovative finance," governments and the world of development aid are experimenting with new mechanisms to support social innovation.

To facilitate this new approach to investing and to coordinate across the necessary agencies and sectors, several governments have developed initiatives to support social innovation markets. For example, the UK government created Big Society Capital (BSC) to support the development of social investment finance intermediaries. Further goals were to increase awareness of and confidence by promoting best practices, sharing information, and improving links with mainstream financial markets [45]. Canada set up Community Economic Development Investment Funds (CEDIFs) to surmount financial hurdles facing local entrepreneurs and to generate economic development in Nova Scotia [46]. In Australia, the country's first Financial System Inquiry (FSI) since 1997 covered impact investing for the first time at the government level in 2014. The 2014 FSI Interim Report mentioned examples such as The GoodStart syndicate, which raised AUD 165 million for 650 ABC Learning day care centers, which it now runs on a not-for-profit basis; mentioned were also the Government of New South Wales launching Australia's first two SIB pilot programs in 2011 [47]. The SIB pilots raised AUD 7 million in capital for UnitingCare Burnside to support families in facilitating their child's return from foster care, and AUD 10 million for the Benevolent Society to prevent family breakdowns [48]. The potential roles that government could play in these and other areas include providing risk capital, creating a dedicated social impact investment bank, as well as tax incentives. All of these were covered in the FSI Interim Report. They provide a good example of early steps a government can take to encourage the development of the impact investing market. The advocacy non-profit Impact Investing Australia accordingly considered the FSI coverage to be "a pivot point for impact investing in Australia" [49].

In spite of all the enthusiasm, the question remains open today how SIBs (and other innovative financing instruments) can unlock their full potential by reaching the scale necessary to be part of an overall domestic and development strategy. Most likely, multiple financial innovations will be needed to provide investment solutions to a wide range of issues that until recently used to fall under the exclusive remit of governments.

14.4 Invest in Virtuous Consumer Innovation

Impact investors investing in advanced economies are very active in the LOHAS segment. LOHAS consumers can be found across a broad array of industries, such as food, fashion, real estate, and transportation. Consumers in the US have built their LOHAS market to over USD 300 billion, growing at more than 10 percent per annum [50]. More than 80 % of US adults are somehow engaged in sustainability, whether they are consciously part of the LOHAS movement or not [51].

While the US consumer has traditionally been viewed as the driver of global growth, this is changing. By 2025, consumption in emerging markets is forecast to have grown by 150 % from 2010 levels, while just by 31 % in developed markets [52]. Emerging markets will be nearly on par with developed markets in trillions of consumption dollars. They will drive global middle-class consumption. The nature of their consumers is different from consumers in developed markets. For example, a lot of the wealth is recent: individuals are five times wealthier than they were just a decade ago [53]. Much of this wealth is simply sitting in savings, ready to be deployed for products and services aligned with the values of consumers (hence the potential to participate in crowdfunding mentioned earlier) [54]. Unlike US consumers who are among the most sensitive to the sometimes-higher prices of green products and services, there is evidence to suggest that emerging market consumers may not be [51]. Data suggests that approximately 84 % of consumers from emerging markets such as China, India, Malaysia, and Singapore would pay a premium for environmentally-friendly products, compared to just 50 % for consumers from the US, Japan, France, and Germany [55]. Accenture warns, though, that, "with the exception of products and services that are niche, clearly offering added value or representing genuine innovation, the market for purposely designed sustainable alternatives will not likely command premium prices in the long run" [56].

Impact investing and the LOHAS segment appear to be a natural fit. Numerous such investment deals are transacted in international impact investor networks such as Toniic, who often back early and growth stage venture capital investments [59]. Established LOHAS topics are organic food, fair trade, and health, with large players including the publicly listed Whole Foods Market, which was discussed in Part I. LOHAS businesses typically source responsibly along the entire supply chain. By building charity into a brand and product offering, some turn a mainstream product such as shoes or glasses into a LOHAS product. For example, TOMS, a US for-profit company with a non-profit subsidiary, Friends of Toms, has pioneered the one-for-one model. Founded in 2006, it designs and sells shoes based on the Argentine Alpargata design and eyewear [57]. For every pair of shoes sold, a pair of shoes is given to an impoverished child; and for every pair of eyewear sold, part of the profit is donated to help restore eyesight to people in developing countries.

Sustainable products will become the new normal in due course. Companies unwilling to adjust their supply chains or even reconfigure their business offerings to prepare for this changing market paradigm are, at best, missing out on an enormous growth opportunity. At worst, they are sealing their fates. As strategist Michael Porter points out, government can be instrumental in the continued competitiveness of these companies: "On the one hand, firms and the private sector are the ultimate engines of innovation. On the other hand, the innovative activities of firms within a country are strongly influenced by national policy and the presence and vitality of public institutions. In other words, innovation intensity depends on an interaction between private sector strategies and public sector policies and institutions. Competitiveness advances when the public and private sectors together promote a favorable environment for innovation" [58]. In the LOHAS space, the most logical policy intervention is to provide for quality health, safety and consumer protection regulation, as well as incentives for sustainable supply chains, rather than necessarily seeding new businesses or intermediaries directly. The next chapter will assess more generally the role governments can play in stimulating impact investing market development.

References

1. OECD (2013) Net ODA disbursements: total DAC countries. http://webnet.oecd.org/dcdgraphs/ODAhistory. Accessed 1 May 2013
2. Prahalad CK (2010) The fortune at the bottom of the pyramid: eradicating poverty through profits. Pierce Education Inc, New Jersey
3. OECD (2013) Aid to poor countries slips further as governments tighten budgets. http://www.oecd.org/dac/stats/aidtopoorcountriesslipsfurtherasgovernmentstightenbudgets.htm. Accessed 1 May 2013
4. Moyo D (2010) Dead aid: why aid is not working and how there is a better way for Africa. Farrar, Straus and Giroux, New York
5. OECD (2005/2008) The Accra Agenda for Action. http://www.oecd.org/development/effectiveness/34428351.pdf. Accessed 1 May 2013
6. International Conference on Financing for Development (2002) Monterrey Consensus of the International Conference on Financing for Development. Final text of agreements and commitments adopted at the International Conference on Financing for Development Monterrey, Mexico, 18–22 March 2002. United Nations Department of Economic and Social Affairs. http://www.un.org/esa/ffd/monterrey/MonterreyConsensus.pdf. Accessed 1 May 2013
7. CNNMoney.com (2013) What will the Global 500 look like in 2021? http://money.cnn.com/galleries/2012/news/companies/1207/gallery.global-500-future.fortune/index.html. Accessed 1 Apr 2013
8. World Business Council for Sustainable Development (2010) Vision 2050: executive summary. http://www.wbcsd.org/pages/edocument/edocumentdetails.aspx?id=219. Accessed 1 May 2013

9. Hockerts K, Morsing M (2008) A literature review on corporate social responsibility in the innovation process. http://www.samfundsansvar.dk/file/318819/a_literature_review_corporate_social_responsibility_innovation_process_september_2008.pdf. Accessed 1 May 2013
10. International Telecommunications Union (2012) Measuring the information society. http://www.itu.int/ITU-D/ict/publications/idi/material/2012/MIS2012_without_Annex_4.pdf. Accessed 1 May 2013
11. World Health Organization (2014) Visual impairments and blindness. http://www.who.int/mediacentre/factsheets/fs282/en. Accessed 1 May 2015
12. World Health Organization (2011) Vision 2020: the right to sight. http://www.who.int/blindness/Vision2020_report.pdf. Accessed 15 May 2014
13. Global Envision (2006) Compassionate manufacturing—Aurolab does business. http://www.globalenvision.org/library/10/943. Accessed 10 May 2014
14. Martin M (2007) F4F—finance for foundations, UBS. http://ssrn.com/abstract=1322388. Accessed 1 Apr 2013
15. Deutsche Bank Corporate Social Responsibility (2013) Eye Fund finances eye care for the poor. https://www.db.com/csr/en/social_investments/eye_fund_finances_eye_care_for_the_poor.html. Accessed 1 May 2013
16. World Health Organization (2009) State of the world's vaccines and immunization. http://www.unicef.org/immunization/files/SOWVI_full_report_english_LR1.pdf. Accessed 10 May 2014
17. World Bank Group (2015) Innovative finance for development solutions: initiatives of the World Bank Group. http://siteresources.worldbank.org/CFPEXT/Resources/IF-for-Development-Solutions.pdf. Accessed 1 May 2015
18. Wood A, Martin M (2006) Market based solutions for financing philanthropy. In: Viewpoints. http://ssrn.com/abstract=98009. Accessed 15 Apr 2013
19. PWC (2011) Counting the cost of carbon: low carbon economy index 2011. http://www.pwc.com/en_GX/gx/low-carbon-economy-index/assets/Low-Carbon-Economy-Index-2011.pdf. Accessed 1 May 2013
20. Bucher K (2014) Green investing report examines ways to reduce the financing cost for clean energy. http://www.weforum.org/news/green-investing-report-examines-ways-reduce-financing-cost-clean-energy?news=page. Accessed 12 May 2015
21. Frankfurt School-UNEP Collaborating Centre for Climate & Sustainable Energy Finance, UNEP & Bloomberg New Energy Finance (2015) Global trends in renewable energy investment 2015. http://fs-unep-centre.org/publications/global-trends-renewable-energy-investment-2015. Accessed 12 May 2015
22. McKinsey & Company (2010) Energy efficiency: a compelling global resource. http://g-bit.com/Search/Data-Points/Details/DataPointId/14236 Accessed 4 May 2014
23. McKinsey & Company (2009) Pathways to an energy and carbon efficient Russia: opportunities to increase energy efficiency and reduce greenhouse gas emissions. http://www.mckinsey.com/client_service/sustainability/latest_thinking/greenhouse_gas_abatement_cost_curves. Accessed 1 May 2013
24. Commission of the European Communities (2006) Green paper: a European strategy for sustainable, competitive and secure energy. http://eur lex.europa.eu/legal-content/EN/TXT/PDF/?uri=CELEX:52006DC0105&from=EN. Accessed 10 Apr 2014
25. Eurostat (2015) Unemployment statistics. http://ec.europa.eu/eurostat/statistics-explained/index.php/Unemployment_statistics. Accessed 10 May 2015
26. Erlanger S (2013) Young, educated and jobless in France. In: New York Times. http://www.nytimes.com/2012/12/03/world/europe/young-and-educated-in-france-find-employment-elusive.html?pagewanted=all&_r=0. Accessed 1 May 2013
27. European Commission (2010) Europe 2020: a strategy for smart, sustainable and inclusive growth. http://ec.europa.eu/eu2020/pdf/COMPLET%20EN%20BARROSO%20%20%20007%20-%20Europe%202020%20-%20EN%20version.pdf. Accessed 13 May 2014

28. European Commission (2015) EU-25 energy and transport outlook to 2030. http://ec.europa.eu/dgs/energy_transport/figures/trends_2030/5_chap4_en.pdf. Accessed 15 May 2015
29. R20 Regions of Climate Action (2015) Finance in action. http://www.regions20.org/finance-in-action. Accessed 15 May 2015
30. Solon O (2011) LEDs can save more electricity than solar can produce. In: Wired. http://www.wired.co.uk/news/archive/2011-10/21/leds-solar-energy-consumption. Accessed 14 Mar 2014
31. Garthwaite J (2011) Tough road to getting LED lights on the streets. National Geographic. http://news.nationalgeographic.com/news/energy/2011/01/110120-led-streetlights-ann-arbor. Accessed 13 Mar 2014
32. Regions of Climate Action (2013) Programs in action. http://regions20.org/programs-action/technology-action/lighting-efficiency-project. Accessed 1 May 2013
33. Martin M (2010a) After Copenhagen: perspectives on energy. In: Viewpoint. http://ssrn.com/abstract=1532825. Accessed 1 May 2013
34. Bloom DE, Canning D (2011) Demographics and development policy. Program on the Global Demography of Aging at Harvard University. http://www.hsph.harvard.edu/pgda/WorkingPapers/2011/PGDA_WP_66.pdf. Accessed 1 May 2013
35. Jackson R et al (2010) The global aging preparedness index. The Center for Strategic and International Studies. http://csis.org/files/publication/101014_GlobalAgingIndex_DL_Jackson_LR.pdf. Accessed 1 May 2013
36. Home Office and Ministry of Justice (2013) Reducing reoffending and improving rehabilitation. https://www.gov.uk/government/policies/reducing-reoffending-and-improving-rehabilitation. Accessed 1 May 2013
37. Ragin L, Palandjian T (2013) Social impact bonds: using impact investment to expand effective social programs. http://www.frbsf.org/publications/community/review/vol9_issue1/social-impact-bonds-impact-investment-expand-effective-social-programs.pdf. Accessed 1 May 2013
38. Martin M (2010b) A funding revolution for public goods: social impact bonds. Geneva
39. Social Finance (2013) The Social Finance register. http://www.socialfinance.org.uk. Accessed 1 May 2013
40. Joblinge (2013) The Joblinge register. http://www.joblinge.de. Accessed 1 May 2013
41. Martin M (2011) Four revolutions in global philanthropy. Impact Economy. http://dx.doi.org/10.2139/ssrn.2209746 Accessed 1 May 2013
42. Trading Economics (2015) Germany youth unemployment rate. http://www.tradingeconomics.com/germany/youth-unemployment-rate. Accessed 2 May 2015
43. Tomkinson E (2014) Social impact bond in Augsburg, Germany. http://emmatomkinson.com/2014/08/20/social-impact-bond-sib-in-augsburg-germany. Accessed 5 May 2015
44. Center for Global Development & Social Finance (2013) Investing in social outcomes: development impact bonds. http://www.cgdev.org/sites/default/files/investing-in-social-outcomes-development-impact-bonds.pdf. Accessed 5 May 2015
45. Big Society Capital (2015) The Big Society Capital register. http://www.bigsocietycapital.com. Accessed 5 May 2015
46. Nova Scotia, Canada (2014) Community Economic Development Investment Funds. http://www.gov.ns.ca/econ/cedif. Accessed 15 Aug 2014
47. Financial System Inquiry, Government of Australia (2014) Impact investment and social impact bonds. http://fsi.gov.au/publications/interim-report/03-funding/impact-investment. Accessed 14 Aug 2014
48. Financial System Inquiry, Government of Australia (2014) Financial system inquiry interim report. http://www.afr.com/rw/2009-2014/AFR/2014/07/14/Photos/95451eb6-0ba5-11e4-b9e0-0f2d6d949f68_FSI_Report_Final_Reduced20140715.pdf. Accessed 4 Mar 2015
49. Bowden A (2014) Australian financial system inquiry flags impact investing for consideration. Impact Investing Australia. http://impactinvestingaustralia.com/social-impact-bonds/australian-financial-system-inquiry-flags-impact-investing-for-consideration. Accessed 15 Aug 2014

50. Onelife Media (2013) LOHAS Market. http://www.effectpartners.com/onelifetour/the-lohas-market. Accessed 1 May 2013
51. Marshall J (2010) LOHAS consumers around the World. Natural Marketing Institute. http://www.lohas.com/sites/default/files/lohasconsumers.pdf. Accessed 1 May 2013
52. Mirae Assets Global Investments (2014) Consumption growth: investing in today's emerging markets. http://investments.miraeasset.us/download_file/65/216. Accessed 3 Mar 2015
53. The Economist (2012) The rich and the rest. http://www.economist.com/node/21564418. Accessed 4 Mar 2014
54. Brandes Investment Partners (2014) Five compelling reasons to allocate to emerging markets. http://www.brandes.com/Documents/Publications/Five%20Compelling%20Reasons%20to%20Allocate%20to%20EM.pdf. Accessed 16 Aug 2014
55. Accenture (2013) Emerging-market consumers also more willing to pay a premium for environmentally friendly consumer products. http://newsroom.accenture.com/article_display.cfm?article_id=4921. Accessed 1 May 2013
56. Accenture (2011) Long-term growth, short-term differentiation and profits from sustainable products and services: a global survey of business executives. http://www.accenture.com/SiteCollectionDocuments/PDF/Accenture-Long-Term-Growth-Short-Term-Differentiation-and-Profits-from-Sustainable-Products-and-Services.pdf. Accessed 1 May 2013
57. TOMS (2013) The Toms register. http://www.toms.com. Accessed 1 May 2013
58. Porter ME, Stern S (2002) National innovative capacity. Harvard Business School. http://www.isc.hbs.edu/Innov_9211.pdf. Accessed 1 May 2013
59. Toniic (2013) The Toniic register. http://toniic.com. Accessed 1 May 2013

When Governments Stimulate Supply and Demand 15

15.1 Supply, Direct, Demand: Three Market Policy Levers

It may seem odd to talk about government when we really are after private sector investment. If you hate governments, you may consider moving to the next chapter right away. Notwithstanding, government can support the emergence of impact-oriented content in the investment industry in three primary ways: by stimulating supply, directing capital and regulating demand.

Government can use "supply development policies" to increase the supply of impact investment capital. These policies can take the form of incentives to invest through co-investing or risk sharing with the government, creating investor requirements for impact investing, or directly providing funding for impact investments or the intermediaries that invest in them. Investing in intermediaries is typically preferred over direct investments. For example, in India, the government mandates that a certain amount be loaned to underserved markets. In the European Union, the European Investment Fund (EIF) launched the Social Impact Accelerator (SIA) in 2013, a pioneering pan-European public-private partnership pilot for impact investing in partnership with Crédit Coopératif and Deutsche Bank [1]. Structured as a fund-of-funds, it mobilized an amount of EUR 6.9 billion of capital for investment in impact funds. SIA provides equity finance to social enterprises that promote social inclusion, provide alternative sources of employment for marginalized social groups, and contribute to growth [1].

The UK's Big Society Capital (BSC) is a compelling example of public sector ingenuity in helping to supply capital. BSC launched a GBP 600 million fund in 2012 [2]. The capital was largely funded by mobilizing unclaimed assets from dormant bank accounts and deploying them in a new way. BSC was set up in the UK in 2011 as an independent financial institution and is regulated by the UK Financial Services Authority. It acts as a social impact investment bank and invests in intermediaries focusing on the UK, rather than the whole of the European Union, as in the case of SIA above.

M. Martin, *Building the Impact Economy*,
DOI 10.1007/978-3-319-25604-7_15

As a supply development policy in the US, the Small Business Association helped to create six equity funds that invest in businesses in low-income communities in 2010 [3]. The US CDFI Loan Fund allocates money and tax credits to community-based organizations in low-income communities across the US that provide financing for the purchase of a first home or starting a business. It works directly on improving recipients' credit ratings or helping them to create a monthly budget [4].

The limited availability of government supplied capital to kick-start a national impact investment industry can be a critical barrier. An interesting proposition is what leading civil society scholar Lester Salamon has dubbed "philanthropication thru privatization." Simply put, it refers to creating massive community-based foundations and philanthropic institutions around the world as a by-product of privatization transactions; they are funded by redirecting a portion of the proceeds [5]. The so-called foundations of banking origin in Italy provide a helpful illustration. In 1990, "Italy converted its network of small, non-profit, and quasi-public, savings banks into joint stock companies, but vested the ownership of the resulting stock in a set of foundations, some of them newly formed and some already in existence as operating arms of the former savings banks." The value of the stock that the foundations were able to sell in 1994 was EUR 24 billion [6]. The endowments continued to increase in value until the financial crisis. By 1998, "the combined assets of Italy's foundations of banking origin exceeded EUR 50 billion. Cariplo and Compagnia di San Paolo, two of the largest of these foundations, have both assets in excess of EUR 9 billion each, or approximately USD 13 billion. This puts both of them ahead of such major U.S. foundations as the Rockefeller Foundation (USD 4.1 billion in assets as of 2013) [7], and the Ford Foundation (America's second largest with USD 11.9 billion in assets as of 2013)" [8]. A foundation landscape was thus seeded in Italy. In principle, it commands the resources to enable activity in impact investing.

Alternatively, government can also "direct capital" to better leverage existing investments for social impact. These policies improve the risk and return features of impact investments. They adjust market prices, and improve transaction efficiency as well as market information, for example by harmonizing impact measurement and reporting standards. In the Netherlands, the Green Funds Scheme set up in 1995 offers a tax credit to investors devoting some capital to environmental initiatives [9]. The EU started to direct funds to climate projects by issuing its "Climate Awareness" bonds in 2010 [10].

Lastly, by "stimulating demand" for impact investment and promoting investment readiness, government can further encourage market growth. By strengthening the capacity of impact investment recipients to absorb capital, demand development policies increase the demand for impact investments. There are several examples of these types of policies from around the world. In the UK, the Investment and Contract Readiness Fund (ICRF) was launched in 2012 with ten million British pounds over three years, providing grants between GBP 50,000 and 150,000 to social enterprises that seek to (1) raise at least GBP 500,000 investment, or (2) bid for public service contracts over one million pounds [11].

New corporate legal forms for social enterprises such as the UK Community Interest Company (CIC) or the US Low Profit Limited Liability Company (L3C) can be another structural demand enabler. These legal forms receive preferential tax treatment for commitments to serving the public good and accepting an asset lock [12]. Compared to a nonprofit, such legal forms increase the transparency and tradability of social enterprises. The CIC in particular is required to conduct annual reporting, which is a major precondition for the development of a viable secondary market for impact investments. Once tradable, such securities would render the impact investment capital market more liquid. This lowers barriers to greater participation.

15.2 Use Structures and the Toolbox

Moreover, some of these new social enterprise legal forms enable layered investing: for example, as an LLC, an L3C can distinguish between its members. This allows different entities to buy memberships with different levels of return and risk. Philanthropists could buy the highest-risk membership of an L3C from the grant component of their portfolios, for example, while a small foundation may buy the lowest-risk membership of the same vehicle from its bond component. As a function of their risk preference, this layering would allow a variety of investors to participate in L3Cs. This can expand the capital pool available (however, L3C takeup among US foundations has been limited; and North Carolina already abolished the L3C again in 2014, after introducing it in 2010). See an overview of the main regulatory and market building blocks below (Fig. 15.1).

15.3 Overcome the Barriers via Incentives

Without the prospect of scale, impact investment is no meaningful pathway to direct financial capital toward the opportunities following from the megatrends discussed earlier. Putting the discussion above in perspective, we cannot—and should not—expect governments to attempt to be too much of a protagonist. In fact, this may be downright counterproductive, as government schemes often miss what citizens want and are prepared to pay for. So how can this actually be achieved? Finding the right balance between public and private initiative is important. Fortunately, impact investment is not the first financial industry to go through a market-building phase. Useful lessons can be drawn from similar past efforts. At this stage perhaps most beneficial is the story of venture capital, the industry that shares many of the features also present in impact investment.

VC—broadly defined—is the capital dedicated to equity or equity-linked investments in privately held growth companies [14]. The first venture capital firm was established in the US in 1946 to invest in companies that sprung up after World War II. As institutions were reluctant to invest, individuals primarily financed this first venture capital fund [14, pp. 24–25]. This structure generally continued until the 1980s, when a clarification in a US Department of Labor classification for

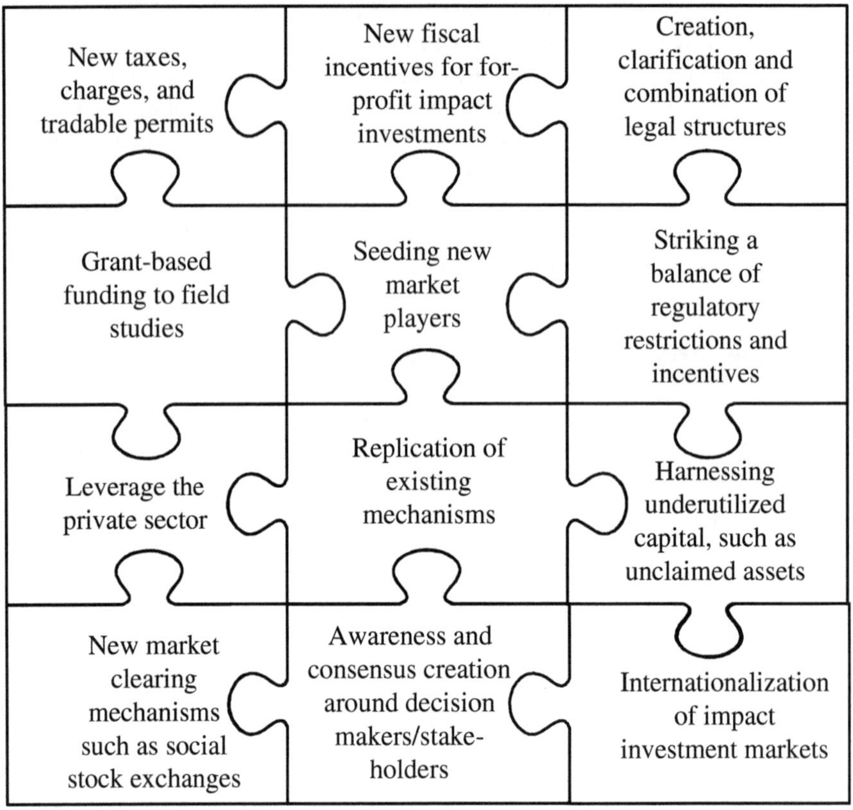

Fig. 15.1 Building blocks for a fertile investment environment. *Source* Author [13]

retirement investments allowed pension funds to invest in venture capital firms as well [14, p. 25].

The US has developed a competitive advantage in venture capital over other countries since the 1980s. Part of this advantage may be due to a robust IPO market. Government policy has played an important role in growing the VC sector [14, p. 33]. Examples like the U.S. Small Business Investment Company (SBIC) and Small Business Innovation Research (SBIR) programs have "disbursed USD 2.4 billion in 1995, more than 60 % of the amount from venture capital in that year" [14, pp. 37–38]. Other countries, such as Germany and Sweden, have replicated the US model and developed substantial government programs to support venture capital [14].

Accordingly, in both the US and Europe government support that encouraged the growth of the venture capital industry took the form of programs that increased the number of investment funds. To achieve this, they adjusted income and capital gains tax policies, addressed employment restrictions, and invested in public R&D. Results indicate that the most effective of these policies tend to be the ones that promote a more active VC industry, not simply those that increase the number of VC funds. Reductions in capital gains taxes have also had a positive effect on VC development. Stock markets for VC-backed companies provide exit opportunities for investors, thus reducing risk. Finally, flexible hiring and firing regulations improve the productivity of the fast-moving VC industry [15].

If impact investing wants to hit it big, it has to provide a superior customer experience, master digital disruption and regulation, and accompany the rise of emerging markets. Incentives can be key to lower the hurdles. An implication from the growth of the VC industry for the impact investment industry is that programs ideally enhance the demand for impact investment funds, rather than simply increasing the supply of capital [16]. Examples of these types of policies include "efforts to facilitate the commercialization of early-stage technology, such as the Bayh-Dole Act of 1980 and the Federal Technology Transfer Act of 1986, both of which eased entrepreneurs' ability to access early-stage research" [16, p. 38]. Similarly, tax policies can incentivize the amount of capital provided and the returns that investments may yield. The lessons from the development of the VC industry and the attempts by different countries to grow their domestic impact investment markets are helpful. This particularly when it comes to addressing the barriers that exist for the industry to graduate to a scale where it can make a major contribution. Included here are: (1) the lack of a set of meaningful rules governing impact investment; (2) limited availability of capital; and (3) a lack of needed skills and expertise and overall investment readiness. Further policies relate to (4) striking the balance between ensuring a level playing field in this new industry in the early phase of an innovation cycle; and (5) the frequent desire to pick winners to fast-track implementation and success.

15.4 Ask Regulators to Enable

Addressing the set of rules governing impact investment is not trivial. In the case of Canada, an early high profile task force was key to charting a path forward. Dedicated to exploring how to better mobilize private and public sector support for impact investing, it surfaced a number of critical challenges inhibiting the growth of the domestic market and proposed a host of corresponding recommendations. While these were not revolutionary insights, the fact that such a task force would deliver comprehensive practical recommendations was instrumental in enabling action. For example, the task force argued that impact investment "has the potential to deliver [USD 30 billion] to public benefit initiatives across Canada" [17]. However, this would only happen if the government were able to create "a regulatory environment that actively encourages the development of investment-ready social enterprises" [17]. The task force also called for Canada's Department of Finance "to amend the Income Tax Act to establish a profits 'destination test' treatment of related business, to serve as the primary regulatory mechanism for social enterprises established and run by charities and non-profit organizations"

[17, p. 22]. Other findings included the need to "review existing legislation and regulations governing charities, non-profits and co-operatives to remove other outstanding barriers to social enterprise activities (e.g. clarifying the asset lock)." Further, the task force made the point that the federal government should take the lead in establishing a consultative process to examine the need for a new class of hybrid corporation, subject to a community interest test, for social enterprises. If appropriate, it should explore options for a "made-in-Canada" model. "This process should leverage the recently launched consultations [...] around a proposed Community Interest Company model" [17, p. 22].

All of this hardly sounds like the equivalent of the solex for capital markets. Yet such solid groundwork is needed to lay the basis for capital market modernization. Governments can do a lot to either enable or disable. When task forces in different member countries share lessons learned, they can accelerate goal achievement in their respective jurisdictions while also preparing the ground for international harmonization. In spite of their inherent risk aversion, committees and task forces are good at the early market development stage. They can create consensus and command the acumen needed to give license to try out something different. Beware though: just as in the case of corporate innovation discussed in Part I, the cases where a committee invented something truly breakthrough in the committee process are—at best—rare.

Looking at market development from another angle, there is a reason why the City of London is vying for the pole position in the innovative Bitcoin virtual currency business: London is Europe's leading financial transactions hub, and has deep expertise in pretty much everything financial. At the other end of the spectrum, a lack of skills and expertise is a barrier inhibiting the growth of the impact investment industry in lesser financial centers. Skills limitations hold back possible investments, raise costs, and lower liquidity.

The foreign exchange markets pose both an interesting opportunity and challenge for impact investors who wish to engage in BoP markets. For example, in 2015, there was still not enough depth in hedging or trading in exotic currencies even in active emerging markets (e.g. the Kenyan Shilling) in the impact investment space. While many impact investments are made into emerging markets, where currency risk is high, most of these investments are made in hard currency. In a 2011 investor survey, 91 % of investments were made in USD, EUR, CAD or GBP, and only 9 % in one of 38 other currencies [18]. To bring the cost of hedging down and increase liquidity, large institutions that have the ingredients of the capabilities in-house need to make them available. In their market-building efforts, governments could incentivize financial institutions to do so, because otherwise initial volumes are too small to support a viable business case.

After so much discussion about the role of government in market development, it is time to look at the other players who are active in the ecosystem of impact investing. Readers who are not fans of government but have stayed with us will appreciate that most of them come from the private sector, and some from civil society. The impact investor ecosystem is the subject of the next chapter.

References

1. European Investment Fund (2013) EIF launches a social impact investing fund of funds. http://www.eif.org/what_we_do/equity/news/2013/social_impact_accelerator.htm. Accessed 14 May 2013
2. Big Society Capital (2013) The Big Society Capital register. http://www.bigsocietycapital.com. Accessed 1 May 2013
3. CRS (2013) SBA new markets venture capital program. https://www.fas.org/sgp/crs/misc/R42565.pdf. Accessed 3 Mar 2014
4. Community Development Financial Institutions Fund (2013) Information for the general public. http://www.cdfifund.gov/what_we_do/need_a_loan.asp. Accessed 1 May 2013
5. Salamon LM (2009) Privatization for the social good: a new avenue for global foundation-building. Privatization Barometer. http://www.feem.it/userfiles/attach/2010525173904PB_Annual_Report_2009.pdf.Accessed 1 May 2013
6. IMF (2014) Reforming the corporate governance of Italian banks. https://www.imf.org/external/pubs/ft/wp/2014/wp14181.pdf. Accessed 6 May 2015
7. Rockefeller Foundation (2012) The annual report 2012. http://annualreport2012.rockefellerfoundation.org/financials/stewardship#sthash.FTJpofA3.dpuf. Accessed 26 May 2015
8. Salamon LM (2009) Privatization for the social good: a new avenue for global foundation-building. http://www.feem.it/userfiles/attach/2010525173904PB_Annual_Report_2009.pdf. Accessed 1 May 2013
9. European Commission (2012) Promoting investment in sustainability: green funds. http://ec.europa.eu/environment/ecoap/about-eco-innovation/business-fundings/netherlands/13112012-promoting-investment-in-sustainability-green-funds_en.htm. Accessed 6 Mar 2014
10. European Investment Bank (2010) Issuance of Climate Awareness Bonds—contributing to expansion of renewable energy and energy efficiency through bond investments of Climate Awareness Bonds. http://www.eib.org/investor_relations/press/2010/2010-067-issuance-of-climate-awareness-bonds.htm?lang=en. Accessed 8 Mar 2014
11. UK Government (2013) Investment and Contract Readiness Fund. http://www.beinvestmentready.org.uk. Accessed 1 May 2013
12. Martin M, Wood A (2008) Unfreezing the foundation asset landscape to create a liquid capital market. http://ssrn.com/abstract=1322308. Accessed 1 Apr 2013
13. Martin M (2013) Making impact investible p 32. Impact Economy. http://www.impacteconomy.com/en/wp4.php. Accessed 9 Mar 2014
14. Bronwyn H, Lerner J (2009) Financing of R&D and Innovation. National Bureau of Economic Research. http://www.nber.org/papers/w15325.pdf?new_window=1. Accessed 1 May 2013
15. Da Rin M et al (2005) Public policy and the creation of active venture capital markets. European Central Bank. http://www.ecb.europa.eu/pub/pdf/scpwps/ecbwp430.pdf. Accessed 1 May 2013
16. Lerner J (2002) Boom and bust in the venture capital industry and the impact on innovation. Federal Reserve Bank of Atlanta. https://www.frbatlanta.org/filelegacydocs/erq402_lerner.pdf. Accessed 1 May 2013
17. Canadian Task Force on Social Finance (2010) Mobilizing private capital for public good. http://socialfinance.ca/uploads/documents/FinalReport_MobilizingPrivateCapitalforPublicGood_30Nov10.pdf. Accessed 1 May 2013
18. Morgan JP (2011) Insight into the impact investment market. http://www.jpmorganchase.com/corporate/socialfinance/document/Insight_into_the_Impact_Investment_Market.pdf. Accessed 8 May 2015

References

16 Growing the Ecosystem of Impact Investment

16.1 Close the Gap

If impact investing is to seize the megatrends described earlier, thrive in an enabling policy environment, and take advantage of the expertise and value chains of large corporations to scale social impact, it needs to mobilize a *huge* amount of capital at all investment stages and across multiple asset classes.

This from a variety of sources. Impact investing furthermore needs to play a relevant role in private and public markets. The good news is that some scale examples already exist, as well as mainstream players embracing impact investing. For example, in 2013, Credit Suisse raised a USD 500 million fund-of-funds to invest in agricultural opportunities in Africa [1]. This is quite an increment when compared to the 2010 Deutsche Bank USD 14.5 million "Eye Fund" for ophthalmological treatment mentioned previously. J.P. Morgan established a Social Finance unit in 2007 that actively co-invests in impact investment funds; and UBS set up a team dedicated to developing innovative products and solutions for its clients as early as 2004 (which I had the privilege to found) [1]. For sure, the industry is still in its infancy. Yet by 2015, more and more banks, asset management and private equity firms were asking how they should position themselves vis-à-vis impact investing, and what skill set and platform they should provide for their clients.

If an influential survey by J.P. Morgan—predicting a market-size profit potential between USD 183 billion to USD 667 billion, and invested capital between USD 400–1,000 billion over the following ten years—is roughly right, we know one thing: we are on to something [2]. The work ahead is substantial though. If we are a bit more accommodating and ask impact investments to grow from USD 60 billion today (as reported by GIIN members) to USD 400–1000 billion in ten years from now, this would still amount to a compound annual growth rate (CAGR) of 21–32 %. If we give impact investing five more years, until 2020—as originally envisioned—, the CAGR rises to 46–76 % to reach USD 400 billion and USD 1 trillion, respectively. To close the gap between aspirations and reality, the next task

M. Martin, *Building the Impact Economy*,
DOI 10.1007/978-3-319-25604-7_16

is then to grow the impact investing ecosystem. A 2013 J.P. Morgan survey of 99 of the world's largest impact investors revealed USD 8 billion of capital committed to impact investment in 2012, and an intention to invest an additional USD 9 billion in 2013 [3, 4]. As reported in the survey's 2015 edition, USD 10.6 billion had been committed in 2014, and another USD 12.2 billion planned for 2015 [5]. It is obvious that the transition path to scale following from the survey is not nearly steep enough—and that one can only hope that the survey suffers from an overly limited sampling of respondents, significantly underestimating actual market size.

Regardless, any attempts to fast-track impact in capital markets need to be grounded in a good understanding of the established division of labor among key players in the investor ecosystem, the focus of this chapter. Crowdfunding platforms, a disruptive force already mentioned in a previous chapter, are discussed in relation to their old economy competitors, financial institutions.

16.2 Use Philanthropic Sector Builders' Advantage

Much of the work to develop the "impact investing industry 1.0"—as we know it today—has been performed by foundation philanthropy. This is where foundations have comparative advantage. In fact, the Ford Foundation pioneered PRI in 1968. Along with grants, PRIs are occasionally used today to help social enterprises in the seed or early stage. However, due to cost and complexity, they are generally passed up [6, 7]. For example, in 2009, only five one-hundredths of one percent of US foundation capital deployed went to equity PRIs [8]. While PRIs have not gained much of a following, they have prompted an important dialogue about earning social returns from giving and investments. In recent years, private investors have been gravitating toward a more business-like approach to philanthropy. They emphasize positive societal outcomes that can be empirically verified, and seek greater accountability in the charitable sector.

Foundations nevertheless continue to play an important role as the impact investing industry matures, especially with respect to disseminating knowledge on best practices and capacity building to get meaningful deal flow started and cut transaction costs. In fact, a well-regulated and well-endowed charitable foundation sector is a major asset in getting the impact investment industry off the ground in any jurisdiction.

16.3 Let Angel and Early Stage Investors Generate Deal Flow

To thrive, impact investing also needs bottom-up innovation and deal flow. This is where angel investors come in with their expertise and "buy and build" strategies. Over the past several decades, angel investors have served as a reliable source of financing for high-growth companies in the US and Europe. The angel investment sector is continually growing and organizing itself through groups and networks. Although venture capital draws the majority of the attention from policy makers, angel investment is actually the primary supply of external seed and early-stage equity financing in many countries. Angel investors also tend to be less sensitive to market cycles than venture capitalists [9]. After the financial crisis, venture capital sources of funds dropped and the businesses they once supported needed financing by angel investors able to fill the equity gap between funding provided by founders and later-stage venture funds [9, p. 9].

Angel investors can bring more to the table than just provide money. They are pivotal for the impact investing industry for five reasons. (1) They can bring expertise in assessing deals—angel investors know how to identify good projects and teams, and how to evaluate them. (2) They can support a wider range of innovation, because angel investors usually invest locally and in a wider range of sectors than venture capital funds. (3) Further, they reduce transaction costs by connecting high-quality entrepreneurs to more investors via groups and networks. (4) They can help build the track record for impact investments—as mentioned, this is a major barrier for institutional investors. (5) Finally, they can support the growth of start-ups in their field of expertise both financially and via business advice [9]. By providing these critical services, angels can bring more impact investments forward to a later stage of financing.

Angel investment does more than helping just the investee to succeed. In the US, estimates suggest angel investors invested approximately USD 24.1 billion in over 76,000 ventures in 2014, resulting in over 264,000 new jobs [10]. Additionally, early stage firms with angel financing have an increased probability of survival and improved performance and growth by 30–50 % on average [11].

For these reasons, greater government support to lower risk for angel impact investment makes sense. Angel impact investing is currently in its formative stage in Europe. I led a first dedicated session and training conducted at the 2013 Annual Conference of the European Business Angels Network (EBAN) [12]. Angel investment framework conditions vary, so policy makers should consider key factors when they draft enabling legislation, such as the level, sophistication, and volume of angel activity in a particular area. Effective policies in one country may not translate well in another. In fact, in the US and Canada, angel policies are implemented at the regional rather than the national level [13].

Tax incentives are a popular policy tool for angel support; but they can be difficult to structure and must be monitored and evaluated to fine-tune [13]. The UK has long-standing angel tax incentive programs that have a demonstrated beneficial effect on the economy. Approximately 24 % of investments would not have been made without one program in particular, the Enterprise Investment Scheme (EIS) [14]. Japan implemented an "income exemption system" in which an angel can deduct from his income the amount of his investment up to a certain limit [15]. In France, the innovation agency Oséo devised a special funding structure for innovative, early-stage companies [16]: the amount of funding is limited to 50 % of the capital needed to draw in other investors. One third of the amount is a grant, and two thirds are a loan. This approach has reduced the risk associated with the development and commercialization of new technologies.

16.4 Woo Professional Investors, the Scale Enablers

Professional investors such as high net worth individuals and family offices have been the backbone of funding impact investing transactions to date, similar to the state of affairs when microfinance opened up for private investment ten years earlier. At the time, eBay founder Pierre Omidyar initiated and funded the Omidyar Tufts Microfinance Fund, a part of the Tufts endowment with USD 100 million earmarked for investments in microfinance [17].

The contribution of professional investors to the industry has been documented in a number of studies [18]. They increasingly want to both "do good" and "do well" with a part of their portfolio. Thus, they look for ethical funds, SRI, and impact investments [19, 20]. SRI, estimated at about USD 3.74 trillion worldwide in 2012, is already at a scale that is relevant; we will discuss active ownership strategies in the next chapter [21]. Driven by personal interest, deep-seated worries about the future of the planet as well as interest in low-volatility opportunities and segments of the market that are uncorrelated with mainstream global benchmarks such as absolute return and capital guarantee products, they help to increasingly drive capital towards impact investments [22].

Because they are scale enablers, it makes sense to woo professional investors. Policy makers can stimulate professional investor engagement by facilitating impact investing products with clear risk-return characteristics, product quality and tax transparency. The case of the EU's EuSEF law illustrates this approach [23]. The EuSEF law is a designation for funds allocating at least 70 % of their capital towards 'social undertakings' via a broad range of financial instruments ranging across the spectrum from equity to debt, with the exception of social undertakings listed on regulated markets or on multilateral trading facilities (MTF). This renders cross-border fundraising quicker and easier. Whether the approach will succeed in significantly contributing to market building will become clear a few years from now. The art lies in striking the right balance between enabling regulation that provides direction and incentives to professional investors, and overregulation that stifles financial innovation because of misguided overprotection of consumers of financial products.

16.5 Go Mainstream via Institutional Investors

A 2013 survey of 47 organizations with GBP 143 billion under management and 4.5 million pension holders revealed that 20 % of respondents thought it was the role of pension funds to invest in impact investments [24]. Highlighting shifting demands placed on institutional investors, 47 % of respondents expected to have some form of impact investments in their portfolio within the next two years. Notwithstanding, institutional investors need scale, return, and liquidity. By 2015, they had not been overly active in impact investing. Rather, they went for socially responsible investing (SRI). SRI refers to an investment concept that seeks to integrate non-financial criteria, encompassing environmental, social and governance (ESG) issues, in its investment decision-making process and portfolio management of publicly listed for-profit companies. The four major approaches commonly used to implement and interpret SRI are positive screening, negative screening, engagement, or integration. A big barrier to institutional impact investing is the discomfort that many investment managers feel in pursuing something other than risk-adjusted financial return, namely social return. The lack of track record does not help either [25].

Notwithstanding, there is growing skepticism among institutional investors that ESG investing is becoming a ticking-the-box exercise. In this view, ESG fails to effectively direct capital to those corporations who are best at innovating to reconcile economic with social and environmental value creation. Moreover, too little ESG relevant information is available on companies from emerging markets, where the growth is. If impact investing can step up product development to better fit institutional needs, it may well find a more receptive audience than anticipated.

One way forward how to make impact investments suitable for institutional investors has been de-risking. This works when a patient player such as a foundation plays an active role in making the solution happen. For example, from 1993 to 2004, hikes in real estate rents in Italy were outstripping increases in family income by four times [26]. This created an urgent need for affordable housing. The Italian government was more accustomed to passing legislation for the vast majority of Italians who own their homes. It had no easily available solution on the shelf [27]. Instead, the Fondazione Cariplo, an Italian foundation, created the Fondazione Housing Sociale (FHS) in 2004 to act as an innovator of housing policy in the Italian real estate sector and to promote the growth and enhancement of social housing. The foundation did not design FHS to simply create living space; its aim was to build an integrated, lively community, complete with services and easy-to-access businesses [28]. In 2006, with initial foundation funding of EUR 85 million, FHS started the first ethical real estate fund, Fondo Abitare Sociale 1, for institutional investors only [28, p. 29]. The Fund's investors accepted a target yield of two percent over inflation, capped at four percent over inflation with any excess going to other social housing initiatives. Importantly, the structure of the Fondo Abitare Sociale 1 was such that it set up a working party to collaborate with the municipal authority and the other sponsors of the initiative. It assigned the role of developer of the social housing project to this multi-faceted group [28]. The fund was very successful in developing units, services, and public benefit structures. In 2008 it caught the attention of the Italian government [28]. Realizing the potential of the model, the government injected further investments. This catalyzed more than EUR 2 billion from banks, insurance companies, and government in a matter of months [28]. As a result, the Fondo Investimenti per l'Abitare (FIA), a national fund-of-funds, was established to manage this capital. Its investment strategy focuses on building social housing units for families "unable to meet their housing needs on the marketplace, but with incomes higher than those which would entitle them to public housing" [28, p. 31]. The foundation's first fund is the model upon which the FIA is based [28].

To gain traction in scale, the impact investing industry needs to systematically turn out products that are suitable to institutional investors. Modest margins and the long lead-time to scale impact investments have thus far made them largely unattractive to institutional investors, not to mention the high risk of failure [29]. However, this is changing. With the advancing discussion on the risk of "stranded assets" of blue chip companies that pension funds typically invest in—investments with large negative externalities that could depress their value in the end, such as in the oil and gas industry—impact investments are bound to become more attractive to this group over time [30]. A major next step to move capital in scale will be to come up with methodologies that go beyond SRI-type screening to invest for impact in public markets for bonds and equity, possibly one of the biggest opportunities in impact investing around the corner.

16.6 Leverage DFIs for Emerging Market Impact Investing

Development finance institutions (DFIs) are increasingly ambitious about impact investing. Their sweet spot is the world of emerging markets. They have capital and clout. Depending on their degree of independence from their respective governments or the lead governments calling the shots in the case of multilateral agencies, they are also more or less politicized organizations, but often thrive on excellent relationships in the regions that they cover.

Bilateral and multilateral DFIs are actively developing their potential for impact through partnerships with other impact investors, often through loan syndications and equity funds. The Dutch development bank Netherlands Development Finance Company (FMO), one of the largest European bilateral private sector development banks with a EUR 8 billion portfolio, communicates enthusiastically about its impact investment activities [31]. FMO seeks to specialize in sectors where its contribution can have the highest long-term impact, such as financial institutions, energy, housing, and agribusiness, food, and water. In 2012, the UK's Department for International Development (DFID) established the GBP 75 million DFID Impact Fund managed by a development finance institution, the CDC, to invest in impact investment intermediaries internationally with a mix of instruments including debt, direct equity investments, and guarantees. The CDC aimed to invest GBP 2 billion from 2011 to 2015, averaging GBP 400 million per year, with debt and direct investments each representing up to 20 % of CDC's portfolio towards 2015 [32]. One ocean away, the IDB Group is the Latin American DFI leader in channeling impact investment capital, particularly to Caribbean and Latin American countries underserved by commercial banks. The extent to which impact investing has become an aspiration in the world of DFIs becomes apparent when considering the IDB Group's statements. It stresses that each of its four private sector windows engages in impact investing and collaborates with impact investors to co-finance deals. (1) The Multilateral Investment Fund (MIF) supports access to finance, markets, and basic services through loans, guarantees, equity and quasi-equity, as well as grants and technical assistance. (2) The Opportunities for the Majority initiative (OMJ) invests in innovative business models that provide high-quality goods and services to low-income people. (3) The Structured and Corporate Finance (SCF) department leads the IDB's non-sovereign guaranteed lending in areas of infrastructure, financial markets, services and industries. (4) Finally, Inter-American Investment Corporation (IIC) provides direct debt and equity financing and technical assistance to small and medium-sized enterprises (SMEs) [33].

16.7 Financial Services Drive Intermediation and Wholesaling

Banks are reeling from fresh waves of regulation and digital disruption. Notwithstanding, leading financial institutions are crucial for any functioning financial market. Some are now starting to embrace impact investment on a global scale and are positioning themselves as potential key actors in order to mainstream this nascent space.

Traditionally concerned with achieving financial returns for clients, major financial institutions are starting to see impact investments as an integral product of their offering for various reasons. First, there is latent client demand. Impact investment products present an adequate response to the rising investor appetite for investment decisions aligned with personal values, calling for a larger and measurable impact. Second, with falling profitability in established asset classes, the impact industry presents an attractive potential for profit over the long term while the achievement of social returns can also boost reputation. Since the financial crisis of 2008, there has been a rise in such financial products in a number of countries. Examples of large banks entering the impact investing space include J.P. Morgan Social Finance, a dedicated unit launched in 2007 to service the growing market for impact investments, or the "Investing with Impact Platform" launched by Morgan Stanley in April 2012 [34]. The key idea is to allow investors to access a range of investments that have been evaluated for their financial return potential as well as societal impact [35]. UBS, an early philanthropic services and impact investing pioneer, launched an impact investing fund-of-funds in 2013 together with Swiss emerging market private equity boutique Obviam [36].

While many financial services firms are promoting their impact investing offering, the reliability and availability of the data is often fairly limited. There are some cases of disconnect between the state of implementation and what is being communicated to the industry and to clients. However, there are also well-established examples that highlight the rise of impact investing. Deutsche Bank has been a pioneer in microfinance and manages over USD 200 million in microfinance assets, and a community development loan and investment portfolio of USD 550 million. Separately, the Deutsche Bank Americas Foundation has an USD 10.1 million PRI portfolio, which is a relevant player in the impact investing industry [37]. The New Initiatives Fund (NIF) of the Deutsche Bank Americas Foundations, established in 1997, provides flexible financing for community development projects that serve the needs of poor and disadvantaged communities within the US and throughout the developing world through a more than USD 16 million PRI portfolio [38]. Another leading actor is Credit Suisse, which managed USD 1.8 billion in impact investments in 2013, primarily focused on microfinance and sustainable agriculture, on behalf of approximately 4000 clients, including private individuals, institutional investors, and development finance institutions via specialist fund manager responsability [39]. Frequently lacking in-house expertise, banks often collaborate with established and specialized investment firms. In January 2012, responsability and Credit Suisse structured and distributed the Fair Trade Fund, which provides working capital to agricultural cooperatives supporting the development potential of small farmers in developing countries [40].

16.8 Banks and Crowdfunding: Kickoff Is When?

Together with the other players discussed above, the financial services industry has a game changing potential to develop and distribute new products and thus channel capital to the real economy as sustainability edges to becoming the new normal of doing business.

Digitization and big data also make it easier to make impact easily measurable at low cost, a precondition for impact investing to become a serious part of mainstream investing. The biggest potential for disruption lies in the reduction of costs and product innovation enabled by digitization on the one hand, and in the rising competition between banks and online P2P platforms. As crowdfunding truly takes off, it will render investments in faraway early stage ventures much easier and cheaper, and impact investing will also benefit. Emblematic crowdfunding organizations are already well established in the impact space. Since its creation in 2005, kiva, a non-profit organization with a mission to connect people through lending to alleviate poverty, has lent over USD 730 million to recipients in 85 countries and mobilized more than 1.3 million lenders [41]. Many more P2P platforms are coming up, and some will succeed. Some have funny names, such as the French platform "Babyloan." Even Germany's poorest city-state, Bremen, now runs a crowdfunding platform called "Schotterweg" (roughly, the 'road to dough' [money]) [42].

It is merely a matter of time until a major financial institution will buy or invest in a P2P lender [43]. Further, the EU has started to work on a crowdfunding framework: its member states desperately need growth and will try out a lot to close an estimated EUR 2 trillion funding gap for small companies in the EU by 2020 [44]. While many of these investments would hardly fulfill the risk and liquidity requirements of an institutional investor today, it is just a matter of time until the digital revolution will have enabled a much more sophisticated funding and trading paradigm. It will be interesting to see if banks will be slow to adapt, as critics assume. Years of punitive regulations and finger-pointing from all directions have understandably rendered the banking industry somewhat gun shy. Yet for impact investing to go mainstream, the boosters are pipeline (investment content), institutional investors, and banks, in addition to the roles already performed by the other players in the ecosystem. May the aircraft carriers turn and decide it is time to get into the game in a bigger way before it is too late. To move from a promising niche to a fully-fledged "post-casino" paradigm, we urgently need more effort to situate impact investing innovation in the overall field of forces that is reshaping finance, and tease out where the growth opportunities are, the focus of the following chapter.

References

1. World Economic Forum and Deloitte Touche Tohmatsu (2013) From the margins to the mainstream: assessment of the impact investment sector and opportunities to engage mainstream investors. http://reports.weforum.org/impact-investment/3-impact-investment-sector-assessment/3-2-impact-investment-ecosystem-the-landscape-today. Accessed 26 May 2015
2. Morgan JP (2010) Impact investments: an emerging asset class. http://www.thegiin.org/knowledge/publication/impact-investments-an-emerging-asset-class. Accessed 26 May 2015
3. Morgan JP (2013) Perspectives on progress: the impact investor survey. http://www.thegiin.org/knowledge/publication/perspectives-on-progress. Accessed 26 May 2015
4. GIIN (2011) Improving livelihoods, removing barriers: investing for impact in mtanga farms. http://www.thegiin.org/knowledge/publication/improving-livelihoods-removing-barriers-investing-for-impact-in-mtanga-farms. Accessed 26 May 2015
5. Saltuk Y, El Idrissi A (2015) Global social finance. Spotlight on the market: the impact investor survey. http://impactinvestingaustralia.com/countriesandregions/uk/j-p-morgan-giin-5th-annual-impact-investor-survey-launched. Accessed 15 May 2015
6. Ford Foundation (2013) Program-related investment. http://www.fordfoundation.org/grants/program-related-investment. Accessed 1 May 2013
7. UNC School of Government (2013) Program-related investments (PRIs): a potential funding source for community economic development? http://ced.sog.unc.edu/?p=4515. Accessed 1 May 2013
8. Bannick M, Goldman P (2012) Gaps in the impact investing capital curve. In: Stanford Social Innovation Review. http://www.ssireview.org/blog/entry/gaps_in_the_impact_investing_capital_curve. Accessed 27 Sept 2012
9. OECD (2011) Financing high-growth firms: the role of angel investors. http://www.eban.org/wp-content/uploads/2013/03/OECD-Angel-Financing-Publication.pdf. Accessed 1 May 2013
10. Center for Venture Research (2013) The angel investor market in Q1Q2 2013: a sustainable growth continues. https://paulcollege.unh.edu/sites/paulcollege.unh.edu/files/Q1Q22013AnalysisReporte.pdf. Accessed 27 May 2015
11. Kerr WR et al (2011) The consequences of entrepreneurial finance. http://www.people.hbs.edu/wkerr/KLS-Angels-Oct2011.pdf. Accessed 27 May 2015
12. Impact Economy (2013) Impact Economy at EBAN's annual congress—Europe's Answer to Silicon Valley. Geneva
13. OECD (2011) Financing high-growth firms: the role of angel investors. http://www.eban.org/wp-content/uploads/2013/03/OECD-Angel-Financing-Publication.pdf. Accessed 27 May 2015

14. Brest P, Born K (2013) Unpacking the impact in impact investing. http://www.ssireview.org/articles/entry/unpacking_the_impact_in_impact_investing. Accessed 5 Mar 2014
15. Morinbu S (2002) Comparison of Japanese and US tax base and changes of tax base in Japan for use at the IMF conference. https://www.mof.go.jp/pri/research/discussion_paper/ron045a.pdf. Accessed 8 May 2013
16. Republique Francaise (2012) Trésor-Economics: the role of Oséo in the financing of innovation. https://www.tresor.economie.gouv.fr/File/372129. Accessed 3 May 2014
17. Chu M, Barley L (2013) Omidyar Network: pioneering impact investment. Harvard case study solution & analysis. Harvard Business School. http://www.harvardsolution.com/omidyar-network-pioneering-impact-investment-589. Accessed 1 May 2013
18. E.T. Jackson and Associates Ltd (2012) Accelerating impact: achievements, challenges and what's next in building the impact investing industry. Rockefeller Foundation. https://assets.rockefellerfoundation.org/app/uploads/20120707215852/Accelerating-Impact-Full-Summary.pdf. Accessed 20 Oct 2015
19. US SIF Foundation (2012) Report on sustainable and responsible investing trends in the United States. http://www.ussif.org/files/Publications/12_Trends_Exec_Summary.pdf. Accessed 1 May 2013
20. Chamberlain M (2013) Socially responsible investing: what you need to know. In: Forbes. http://www.forbes.com/sites/feeonlyplanner/2013/04/24/socially-responsible-investing-what-you-need-to-know. Accessed 24 Apr 2013
21. The Forum for Sustainable and Responsible Investment (2012) Sustainable and responsible investing trends in the United States. http://www.ussif.org/files/Publications/12_Trends_Exec_Summary.pdf. Accessed 23 May 2015
22. The Economist (2010) The new world of wealth: seven key trends for investing, giving and spending among the very rich. http://evpa.eu.com/wp-content/uploads/2010/10/The-new-world-of-wealth-Seven-key-trends-for-investing-giving-and-spending-among-the-very-rich.pdf. Accessed 1 May 2013
23. European Securities and Markets Authority (2015) ESMA's technical advice to the European Commission on the delegated acts of the regulations on European social entrepreneurship funds and European venture capital funds. http://www.esma.europa.eu/system/files/2015-esma-227_-_final_report_on_advice_on_eusef-euveca.pdf. Accessed 9 May 2015
24. World Economic Forum and Deloitte Touche Tohmatsu (2013) From the margins to the mainstream: assessment of the impact investment sector and opportunities to engage mainstream investors. http://reports.weforum.org/impact-investment/3-impact-investment-sector-assessment/3-2-impact-investment-ecosystem-the-landscape-today. Accessed 26 May 2015
25. Social Finance and Finetic (2012) Microfinance, impact investing, and pension fund investment policy survey. http://www.socialfinance.org.uk/sites/default/files/final_pension_fund_survey_october_2012.pdf. Accessed 1 May 2013
26. Bank of International Settlements (2015) Housing finance in Italy. https://www.bis.org/publ/wgpapers/cgfs26gobbi.pdf. Accessed 5 May 2015
27. Fondazione Housing Sociale (2014) A foundation for social housing in Italy. http://www.fhs.it/eng.pdf. Accessed 1 May 2015
28. European Commission (2014) Fondazione Housing Sociale: getting the market to target public needs. https://webgate.ec.europa.eu/socialinnovationeurope/directory/news/fondazione-housing-sociale-getting-market-target-public-needs. Accessed 1 May 2014
29. Koh H et al (2012) from blueprint to scale: a case for philanthropy in impact investing. Monitor. http://www.thegiin.org/knowledge/publication/from-blueprint-to-scale-the-case-for-philanthropy-in-impact-investing. Accessed 1 May 2013
30. Carbon Tracker Initiative (2013) The Carbon Tracker register. http://www.carbontracker.org. Accessed 1 May 2013

31. FMO (2015) Results FMO underline increasing development impact. https://www.fmo.nl/k/news/view/18455/538/results-fmo-underline-increasing-development-impact.html. Accessed 15 Apr 2015
32. CDC (2013) DFID Impact Fund. http://www.cdcgroup.com/dfid-impact-fund.aspx. Accessed 1 May 2013
33. IDB (2013) Impact investing in Latin America and the Caribbean with the IDB Group. http://www.iadb.org/en/resources-for-businesses/impact-investing-in-latin-america-and-the-caribbean,5752.html. Accessed 1 May 2013
34. J.P. Morgan Chase and Co. (2013) J.P. Morgan Social Finance. http://www.jpmorganchase.com/corporate/socialfinance/social-finance.htm. Accessed 1 May 2013
35. Morgan Stanley (2013) Investing with impact. http://www.morganstanley.com/globalcitizen/investing-impact.html. Accessed 1 May 2013
36. Obviam (2013) Investing for impact. http://www.obviam.ch. Accessed 1 May 2013
37. Global Impact Investing Network (2015) Investor's council members. http://www.thegiin.org/cgi-bin/iowa/council/member/index.html. Accessed 7 May 2015
38. Deutsche Bank (1996) The New Initiatives Fund: a portfolio of social investments. https://www.db.com/us/docs/New_Initiatives_Fund.pdf. Accessed 20 Oct 2015
39. GIIN (2013) Investors' council members. http://www.thegiin.org/cgi-bin/iowa/council/member/index.html#73. Accessed 1 May 2013
40. responsibility (2012) First global fair trade fund launched. http://www.responsability.com/domains%5Cresponsability_ch/data/secure_pdf/20120112_fairtradefund_e.pdf. Accessed 1 May 2013
41. Kiva (2015) The kiva register. http://www.kiva.org/about. Accessed 15 Jul 2015
42. City of Bremen's (2015) The city of Bremen's register. https://www.bremen.de/schotterweg. Accessed 15 Jul 2015
43. Alois JD (2015) Crowdfunding predictions for 2015. In: Crowdfund Insider. http://www.crowdfundinsider.com/2015/01/60503-crowdfunding-predictions-2015. Accessed 15 Jul 2015
44. Alois JD (2014) Europe faces a funding gap of €2 trillion by 2020, can investment crowdfunding fill the void? (Deck). http://www.crowdfundinsider.com/2014/12/59544-europe-faces-funding-gap-eur-2-trillion-2020-can-investment-crowdfunding-fill-void-deck. Accessed 15 Jul 2015

Toward a Post-casino Paradigm 17

17.1 Vote that Proxy

The world's largest asset manager BlackRock declared 2015 the year of divergence, with tighter financial conditions in the Eurozone and a monster bet on monetary stimulus in Japan, as well as divergent economic growth patterns and a number of zero-sum conflicts such as the geopolitical confrontation over Ukraine [1]. Perhaps underappreciated, another red thread is the enormous value creation opportunity resulting from the sustainability megatrends discussed earlier. If we master the *how-to* questions discussed throughout this book, the world need not converge on the overshoot and collapse scenario outlined in *Limits to Growth* that—on our current path—though behind schedule, looks like the ultimately likeliest outcome [2].

In 2014, BlackRock engaged with 1500 public companies, voted clients' shares in 94 markets and managed USD 252 billion in specific ESG mandates [3]. To do so, the firm used a combination of proprietary methodology and third-party research to assess portfolio companies' ESG performance and risk exposure. Enabling this approach is shareholders' right to vote on official company affairs. By attending a company's annual general meeting, shareholders can do so either directly—or indirectly, by submitting a proxy ballot. Public corporations often have very large numbers of shareholders. As only a minority attends the annual meeting, most voting takes place by proxy. Such proxy ballots provide shareholders opportunities to influence company policies and practices. The shareholder resolution has perhaps been the most frequent practice in the US since the 1940s [4]. When shareholders file resolutions, they often want management to engage with them face-to-face; in fact, once such meetings take place, resolutions are frequently withdrawn [4].

In the past 20 years, proxy voting has become an important tool of active ownership. In 1998, it became a responsibility for pension plan fiduciaries in the US; proxy voting is common practice in the EU as well [5]. The goal was to make sure that the interests of plan participants in their retirement income would not be subordinated to unrelated company objectives and that shareholder value would be duly pursued. Given their long-term focus, pension funds care about future value

M. Martin, *Building the Impact Economy*,
DOI 10.1007/978-3-319-25604-7_17

beyond a few quarters. Unsurprisingly, proxy voting now increasingly incorporates the assessment of environmental and social practices: they have long-term economic implications. As sustainability becomes integral to value creation, their relevance is bound to rise. Active owners accordingly frequently ask the company to prepare a report on environmental or social issues that affect some aspect of company operations.

There is a problem though. Over time, certain governance principles have become widely accepted as representing "best practices." These include sound corporate governance principles, such as a board of directors that has term limits, is run by an independent chairperson, offers diversity in terms of board representation, has independent audit, nominating and compensation committees, and elects its directors by confidential, majority vote with equal voting rights for all shareholders. Other such best practices include the independence of auditors, or linking executive compensation to company performance. For example, in the US, the SEC requires company proxy statements to include information on [6]:

- Compensation policies and their relationship to risk management;
- Qualifications of director nominees;
- Legal actions involving company executives, directors and director nominees;
- The role played by diversity in the consideration of director nominees;
- A board's leadership structure, including whether chairperson of the board and the chief executive officer (CEO) are two separate individuals, and if not, why not;
- Stock awards to company executives; and
- Any conflicts of interest involving compensation of consultants.

While proxy voting has led to sophisticated corporate governance practices, the progress has been more humbling with respect to long-term sustainability. Successful shareholder resolutions often ask companies to prepare sustainability reports, looking into topics such as energy efficiency, CO_2 emissions, or water usage. They then disclose to shareholders the actions taken to mitigate risk. A second line of action are shareholder resolutions that call upon companies to endorse or operate in conformity with widely accepted international standards of conduct such as the United Nation's Universal Declaration of Human Rights, the United Nation's Global Compact, or the International Labour Organization's Declaration on Fundamental Principles and Rights at Work. Many corporations choose to address the ensemble of these issues by adopting sustainability reporting. One leading approach is the one developed by the Global Reporting Initiative (GRI). GRI is an international standards organization with the mission to assist businesses, governments and nonprofit organizations to assess as well as communicate the impacts of their operations on major issues including climate change, corruption, or human rights [7].

These resolutions are generally laudable. Notwithstanding, the problem is that such a compliance and policing logic is not automatically conducive to embracing the opportunities of business transformation that were discussed in Parts I, II, and III. One key opportunity for impact investing therefore consists of taking proxy voting up a notch. This means mainstreaming the idea that companies should engage in activities and adopt structures that enable them to systematically seize the business potential associated with sustainability megatrends such as the ones discussed earlier. These include achieving adequate exposure to emerging markets in their industry and to new technologies that enable achievement of total resource productivity goals. Not only could this steer a substantial volume of assets into a productive direction, it would also make businesses' deep industry expertise much more widely available to earlier stage ventures that stand at the beginning of the CIV ladder.

17.2 Drive Impact Where the Action Is

When we look for innovative investment content at a large scale, let's not forget that value is created where the action is. Consider advancing Africa, for many the next big thing, for others already a sobering experience. In 2008, Africa counted one billion people with USD 1.6 trillion GDP and USD 860 billion in combined consumer spending; foreign direct investment in 2010 was USD 55 billion [8]. The 2020 forecast is a GDP of USD 2.6 trillion, and roughly USD 1.4 trillion in combined consumer spending [8]. In 2040, when Europe and China will have greyed, we can expect 1.1 billion Africans of working age [8].

If the current trends continue, Africa will play an increasingly important role in the global economy more generally. The continent can be an important beneficiary of changes underway in the world of finance as well, provided it offers an attractive operating and financing environment. This is where financial innovation and financial engineering can help. Pent-up demand will drive growth in Africa for decades to come. Only 19 % of the African population has access to the Internet [9]. Seven out of ten people living in sub-Saharan Africa do not have access to electricity [10]. With abundant resources and fast-growing market opportunities, the continent's potential is undisputed. Devising solutions that simultaneously respond to nonlinear challenges such as global warming will be key for driving long-term prosperity. Today, consumer-facing industries on the continent already grow two to three times faster than in the OECD [11]. For Eastern and Western companies, early entry into African economies can provide opportunities to create markets, establish brands, shape industries, influence consumer preferences, and foster long-term relationships.

Fully seizing the opportunity will mean empowering people on a large scale and ensuring inclusive growth. Massive capital investment is needed. Alongside ensuring macroeconomic stability, governments can also help by creating the framework conditions for capital to flow. Next to a facilitating framework for equity investments, this also includes the presence of local banks who are able to make loans to emerging private sector activities, and channel longer-term capital.

There are few places where the ways this conundrum will be resolved are more important than in Ethiopia whose potential to develop a fully-fledged textile and garment industry we discussed already. Ethiopia is the second-poorest country in the world, with the fifth-largest population of poor people in the world after India, China, Bangladesh and Pakistan. The 2014 global Multidimensional Poverty Index (MPI) published by Oxford University classifies 87.3 % of Ethiopians as "multi-dimensionally" poor and 58.1 % as destitute; the country was better ranked only than Niger [12].

Ethiopia is however in the fortunate position to be a strategic ally of the main donor governments, with corresponding external budget support, and benefits from joint donor and private remittances that were 70 % greater than the country's total export earnings in 2014 [13]. The National Bank of Ethiopia (NBE) achieved a major win by cutting inflation from 33 % in 2011 to single digits in 2014 [13]. Moreover, the country's banking sector was well capitalized with a capital adequacy sector ratio (CAR, i.e., a bank's tier one and tier two capital divided by its risk weighted assets such as credit exposures) well above the 8 % mandated minimum, with non-performing loans in the 2 to 3 % range [13].

To take advantage of the underlying opportunity and innovations underway in capital markets, for countries such as Ethiopia, the next step is to find ways to initiate the transition from a state-led development model to one where the private sector plays a larger role. This within a general frame of sustainable development. Included here is the need to locate ways to cover government deficits beyond central bank financing, and to develop an attractive government bond market. To matter, impact investing needs to play where the action is, and use financial innovation to properly price risk and allocate capital in theaters where mainstream finance has found it difficult to play in the past. Technology driving transparency and layered capital structures enabling de-risking for different types of investors can both be useful. Just like in renewable energy or garment manufacturing, things need to work in practice in impact investing as well. Navigating the transformation of capital markets can yield a handsome benefit in the different sectors of the real economy. For example, at the time of writing, Ethiopia was offering over eight million acres of land to commercial farming investors, promoted industrial sectors and provided basic infrastructure to develop four industrial cluster zones [14]. To overcome multi-dimensional poverty, significant progress in education, health and living standards is needed. This can only be achieved on the back of a modernizing and competitive economy, and this requires investment.

In developing countries, aligning investments to a larger extent with national development plans and key growth sectors provides an interesting opportunity. Let's have a quick look at two countries that are part of a new set of key countries I introduced as "EMICs" in 2014 [15]. They present exceptional opportunities to achieve both impact and value. The EMICs (Ethiopia, Myanmar, Iran, and Colombia) are high-stakes, high-opportunity countries whose characteristics allow impact investors to make a key difference in enabling them to prosper. Yet all EMICs face important infrastructural and institutional deficits, and raise the question how a responsible and forward-looking approach to investment and business innovation can drive large-scale positive impact. Their success or lack of success to modernize could have more wide-ranging implications. To unpack the argument, here is a further look at Ethiopia.

In 2010, Ethiopia launched its most recent national development plan, the Growth and Transformation Plan (GTP) 2010/2011 to 2014/2015 [16]. The GTP's overarching goal was to reduce poverty and transform Ethiopia into a middle-income country by 2030 or earlier. To achieve this vision, the GTP focused on four major objectives, as well as seven strategic pillars for action (Table 17.1). In addition, the plan attempts to achieve all United Nations (UN) Millennium Development Goals (MDGs).

Key growth sectors for the GTP are agriculture, infrastructure, industry and services. When the plan was adopted, agriculture was Ethiopia's most important industry, accounting for over 47 % of GDP and over 85 % of national employment [17]. The GTP set the goal to grow agricultural industries by 8 % annually. Over the past decade, agricultural industries nearly met this target, growing 7.3 % annually. However, some of Ethiopia's national territory is located along the Sahel, the Sahara Desert's border region. This makes it vulnerable to severe droughts and irregular climate events [18]. The last severe drought occurred in 2004 and placed nearly 20 million Ethiopians at the risk of famine, while also leading to a contraction of the agricultural industry by over 10 % [19].

Similar to agriculture, Ethiopia also experienced robust growth in industry and services, growing annually by an average of 11 and 12.7 %, respectively, from 2004 to 2013. Altogether, measured from a 2005 baseline, Ethiopia experienced steady GDP growth of 8–12 % annually in current US dollars. By helping to induce such growth, the GTP largely achieved its objectives. In terms of meeting MDGs, Ethiopia also made respectable progress towards the eight goals. The national poverty rate decreased from 38.9 to 29.6 % between 2005 and 2012, close to the MDG target of 24 %. In total, Ethiopia bordered on achieving six of the eight MDGs, excluding maternal mortality and gender equality goals [20]. However, rather than by markets, growth was driven by ultimately unsustainable economic policies partially financed by printing money. Inflation shot up to 30-year record highs, expanding from less than 4 % in 2004, to 33 % in 2011 [21].

Table 17.1 Ethiopia growth and transformation plan goals

Major objectives	Strategic pillars
Annual GDP growth of 11 %	Sustain rapid and equitable growth
Improve provision of health and education	Strengthen agriculture
Create a stable and democratic country	Expand industrial sectors
Improve macroeconomic stability	Develop infrastructure
	Social development
	Good governance
	Empowerment

Source Compiled from The Federal Democratic Republic of Ethiopia [16]

When we ask what kinds of investments could drive impact and profit, it is interesting to briefly compare the Ethiopian experience with that of Myanmar, also a developing country that similarly has a massive need for foreign investment to help drive the country's sustainable development, and already discussed in Part II. Myanmar's overarching development agenda, planned out in the 2012 Framework for Economic and Social Reform, was to integrate with the international community by 2020, while transforming the country into a modern economy [22]. The four major policy levers the government used to achieve these goals were the support of industrial development, foreign and local investment, social development, and the improvement of countrywide statistical data.

As of 2014, national data on the social, environmental and economic conditions of Myanmar was profoundly deficient, when compared with most nations. Countrywide macroeconomic targets for the 2012–2015 period included sustaining an annual GDP growth rate of 7.7 %, growing the industrial sector to 32 % of GDP and increasing per capita income GDP 30–40 % by 2015 [22]. Similar to the Ethiopian economy, Myanmar was heavily reliant on a rural and largely agricultural workforce. The national development agendas for the two countries were different, as the Myanmar plan sought to gradually move the country away from a reliance on agriculture, placing a stronger emphasis on value added services and industry. The Ethiopian plan looked to incorporate high value added agricultural products, like floriculture and horticultural products for export. However, the two countries have followed a remarkably similar path in expanding cereal production over time (Fig. 17.1). Key sector targets for Myanmar's economic growth were natural resource exports, infrastructure development, telecommunications development and building domestic energy infrastructures [23]. One key target for telecommunications was to expand mobile coverage density from 10 to 80 % by 2015 [24].

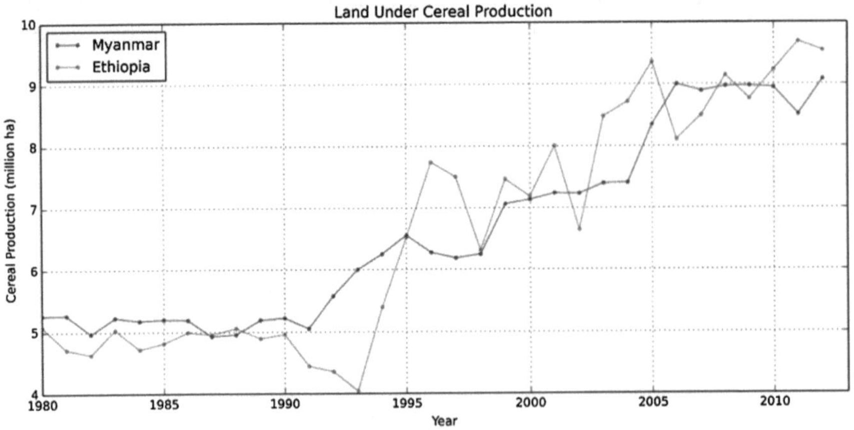

Fig. 17.1 Ethiopia and Myanmar cereal production. *Source* Calculated from World Bank [25]

As well as making an important difference, finding ways to invest in scale and for impact in emerging countries would provide some of the fresh investment content needed. Just consider that Iran has a young population, but no IKEA. Or that in Cuba, 70 % of the housing stock is deficient, but mega do-it-yourself stores do not exist, and that the country's light manufacturing industry is confronted with technology obsolescence. Countries such as Ethiopia and Myanmar will be more difficult investment propositions, but greater alignment with national development plans and the use of modern financial engineering could help. Designing products that appropriately link risk and reward in the parts of the world where the action is could go a long way to enable impact investing on a much larger scale.

17.3 Invest for Impact in the Internet of Things

Capital markets are not just disrupted in spatial terms by the rise of emerging markets. They are also being changed in profound ways by the digital revolution, as we already discussed. But here is the twist: so is the rest of the economy. This provides very substantial fresh opportunities to invest for impact. During the twentieth century, many observed that developments in computer hardware and information technology increased at exponential rates. Gordon Moore of Intel observed that the number of transistors per microchip doubled every 18 months [26]. Martin Cooper, the "father of the cell phone," observed that the total bit capacity of wireless communication seemed to double every 30 months [27]. As more and more people utilize computers and smart phones, this exponential growth has also been observed in Internet connected devices. In 1999, Kevin Ashton, a radio-frequency identification (RFID) researcher, coined the term "Internet of Things" (IoT) to describe the use of networked data sensors in retail, manufacturing, healthcare, consumer products and other sectors [28]. Such sensors can gather and share data on device use and ambient conditions. Using sensors to quantify how tools are used can enable machines and people to better understand their environment. CISCO Internet Business Systems Group estimated that the Internet of Things was "born" between 2008 and 2009, when the number of connected devices and sensors surpassed the global human population [29]. In 2013, over five billion wireless chips were produced globally and by 2020, it is projected that there may be over 50 billion interconnected devices, with nearly seven devices per capita [29]. According to CISCO, by the mid-2020s USD 19 trillion in value will be captured from Internet of Things applications [30]. In 2012, McKinsey & Company accordingly argued that capturing this value by increased data collection and analysis represents the single largest opportunity for economic growth and wealth creation [31].

Drivers of this value touch every industry and include better supply chain and logistics, asset utilization, employee productivities and improved customer experiences, as was apparent in some of the product examples provided earlier in Parts II and III of this book. Embedded sensors can rapidly measure light, temperature, pressure, vibration, stress, airflow, humidity, pressure and movement at high

Table 17.2 Select IoT applications

Business	Society	Environment
Supply chain management	Medical	Smart cities
Logistics	Healthcare	Smart water
Aviation	Home and office	Smart metering
Automotive	Media and entertainment	Natural disaster monitoring
Production lines	Smart buildings	Recycling and waste management
		Resource production

Source Compiled from Leary and Kaczmarska [37]

degrees of sensitivity [32–34]. For example, a complete picture of motion can be derived from only three cheap tools: an accelerometer, a magnetometer and a gyroscope [35]. The Internet of Things allows sharing this sensor information anytime, anywhere and with anyone [36]. Unsurprisingly, the quantity and diversity of data collected has a wide range of applications for industry, the environment and society [36]. This is relevant for capital markets and impact investing. Table 17.2 illustrates potential industries and impact for business, society, and the environment.

There is nothing utopian about applying Internet of Things techniques to business operations. For example, in 2014, CISCO modeled increased sensor use in three private sector development projects in South East Asia: oil and gas, utilities and transportation [38]. By installing sensors along a 138 km metro rail line, with a combination of street, monorail and bus services, a USD 5 million investment saved USD 12.2 million annually. Employees made better business decisions regarding track maintenance, energy consumption, worker productivity, management costs and safety. The utilities model produced a ten million dollar annual benefit in energy savings for over five million customers, following from a USD 8.7 million total investment, while the oil and gas model increased employee productivity by 60 %.

At the state level, Internet of Things technologies are also already influencing basic services, infrastructure, health and education. Cities and the public sector have begun to target Internet of Things value, due to challenges associated with job creation, economic growth, environmental sustainability and social issues [39]. In the 2010s, Amsterdam developed a smart public lighting program to reduce costs. Chicago enacted a digital neighborhood services platform to reduce crime. New York began to introduce Wi-Fi platforms and information services throughout the city. Busan developed the "Busan Information Highway" to strengthen connections between universities, people and the metropolitan government [40].

Embracing Internet of Things technologies is especially conducive to value creation in a few key areas. These include asset utilization, employee productivity, supply chain and logistics, customer experience, and time-to-market [41]. Low-hanging fruits for Internet of Things applications include development of smart-grid systems for utilities, smart buildings, connection healthcare and patient monitoring, smart factories, education, commercial vehicles and ground transportation, marketing and advertising and entertainment [41]. As the Internet of

Things takes off, the question is what this means for impact investing. Vastly greater resource and energy efficiency are topics impact investors care about. To which extent should a company's forward looking engagement with such technologies and a commitment to share their benefits widely influence a shareholder's capital allocation to that company?

Taking the argument one-step further, accelerating returns have been observed in genetics, nanotechnology, robotics, artificial intelligence and energy production [42]. For example, since 1977, the average cost of a photovoltaic panel has decreased 20 % per year [43]. Compounding this global decrease, the total installed capacity of solar photovoltaics has increased 40 % per year since 2000 [44]. The declining cost curve of many emerging technologies may serve to democratize the use of goods and services, if end-user costs continue to decrease [45]. Business strategist and futurist Jeremy Rifkin believes that technology development in competitive markets will ultimately reduce the marginal cost of production for most goods and services to near zero [46]. This would mean that the fixed costs of building an Internet-based business can still be relatively high, but the marginal cost of sharing goods and services over the Internet is near zero due to developed Internet infrastructure. Economists J. Bradford DeLong and Lawrence Summers similarly noted in 2002 that the creation of knowledge is cumulative and shared, which over time may help to decentralize the information economy because of the cheap ability to distribute products [47]. In light of these trends—accelerating returns, decentralization, reduction in marginal costs, the expansion of Internet-connected devices—future investment plans must take a dynamic perspective or otherwise risk being overtaken by reality.

To get a sense of possible forward-looking impact investment themes that could help to further grow impact investing to a significant market size and give the energy transition a hand, it is instructive to look at the work Jeremy Rifkin has conducted since 2009 for the metropolitan areas of Utrecht, Rome, San Antonio, Monaco and Nord-Pas-De-Calais. He developed master plans that combine economic, technological, social and environmental areas. Each master plan is framed within the context that historical economic growth is accelerated by the synergies of communication technologies (e.g. printing press, telegraph, newspapers, radio, television and the Internet) and energy technologies (manpower, biomass, fossil fuels, nuclear, and renewables) [48]. The main argument is that *jointly* developing these two infrastructures is of critical importance. According to Rifkin, the communications/energy regime for the twenty-first century will be the combination of Internet connectivity and cheap renewable energy. In consequence, Rifkin's master plans center around five policy levers: developing renewable energy sources, building smart energy infrastructure grids, converting buildings to micro-power plants, developing smart transportation systems and building an Internet of Things infrastructure [48]. Supplementing these levers are energy efficiency measures that unite different areas.

From a capital markets and impact perspective, we need to ask how such infrastructure could actually be financed in ways that make economic sense. Table 17.3 illustrates the host of policies that Nord-Pas-de-Calais, a region in the North of France bordering with Belgium and inhabited by four million people, has begun to enact. Consider the amounts of capital that can be deployed for impact and return once this kind of analysis is rendered compatible with the investment needs and impact goals of private equity funds and institutional investors. Public money could be used to properly de-risk enabling investments that cannot offer the returns and liquidity needed on a standalone basis, but are economically viable and consciously connect the possibilities offered by technological progress with serving social needs and achieving a better environmental footprint.

Table 17.3 IoT policies in Nord-Pas-de-Calais

Energy efficiency	Energy infrastructure	Renewable energy
Youth employment and training	Storage: hydrogen, hydraulic, compressed air	New investment models
Collecting building data	Develop hydrogen to methane production capacity	Renewable export strategy
Citizen participation and stakeholder consultation	Open data sharing services	Consultation and political support
National regulation	Smart metering	Solar data map
Carbon programs and energy efficiency offsetting	Detailed inventory and renovation targets	–
Energy tax	–	–
Promote self-consumption	–	– –
Buildings	Internet of Things	Transportation
Exemplary project in nonresidential buildings	Aim for energy autonomy	Create a transportation planning council
Detailed inventory and renovation targets	Long-term investment plan	Create a mobility agency
Capacity building	Change company structure: shift CIO to chief productivity officer (information, energy, logistics)	Establish a green port
Regional training program	Support smart grid projects	Shift transportation to rail
Building retrofit program	Develop a logistics commons	Modernize and upgrade freight trains
–	Support a multimodal logistics brokering exchange to extend logistics	–

Source Compiled from Rifkin [48]

17.4 To Get to Impact Investing 2.0, Favor "Impact Light"

Engaging in promising theaters in terms of investment content is a step in the right direction. If we want capital markets to resource the sustainability opportunities that are now emerging, we need to develop a practical approach to impact measurement, the differentiator of impact investing.

Interestingly, the impact measurement scene is moving to advocating for a more case-by-case approach to measuring impact, rather than asking for the chief cook and bottle washer. Others argue that "impact management" is at least as important as impact measurement [49]. Given the complications, this is understandable. There is one problem though: how can this help us to attain scale? What's really needed is "impact light:" to translate potential into large scale, impact investing will of course need to engage in market building facilitated by digital technology, smart policy-making and actions to strengthen the impact investor ecosystem. That won't be enough though. Impact investing has to come to terms with the need for increased transparency and forms of social impact measurement that are able to achieve industry consensus. Disclosure of reliable, credible, and comparable performance data is as critical a dimension in impact investment as it is in traditional investment. However, it would be a mistake to set the reporting hurdle so high that only very few investments can pass it if reporting is not subsidized by philanthropic resources.

If we are ambitious, mainstreaming impact investment is possible; the development of a far more effective and larger marketplace could be around the corner. While the advancements achieved thus far are impressive, a key blocker could nevertheless stand in the way of scale. If we cannot sort out the differentiator, we cannot go mainstream. At the most basic level, "investment capital moves based upon its understanding of future opportunity—the risk, reward and performance offered by any given investment, fund or manager" [50]. Disclosure of reliable, credible, and comparable performance data is a critical dimension of traditional investment. Similar transparency must become a part of impact investment, because insufficient information hampers the pricing of risk and impact, and holds back capital allocation.

One central element in increasing the potential attractiveness of impact investing to investors is to increase awareness about what sets impact investments apart from traditional investments: the creation of positive impact. If this impact creation is not demonstrable, the entire concept of impact investing becomes moot. Indeed, the persistent lack of universally applicable and accepted impact measurement standards is considered as one of the top issues facing impact investing. To that end, efforts both direct and indirect of the impact investment industry have attempted to increase transparency via the creation of a few notable, leading measurement systems. Direct systems are the Impact Reporting and Investment Standards (IRIS) and the Global Impact Investment Rating System (GIIRS); and indirect systems include ESG standards, UN Global Compact, UN PRI and the Global Reporting Initiative (GRI).

The systems mentioned above show that the impact investing industry in general and impact measurement specifically have been characterized by the fast-paced development of a set of fragmented standards: too many standards, too little quality! This is reminiscent of the plethora of environmental and social performance labels in the fashion industry. These individual systems have made great strides in creating the ability for investors to measure the nonfinancial performance of their investments. In some cases, they provide the opportunity to compare the relative nonfinancial performance of one investment to another investment, or aggregate benchmark of investments. However, these systems are limited. Unsurprisingly, in the US "a number of foundations have expressed concerns about the applicability of the IRIS indicators to its portfolio of domestic PRIs. In addition, some foundations are questioning whether the types of measures that can be standardized into common indicators (e.g., acres of land, number of seats in schools) are sufficient to demonstrate social impact" [51]. A March 2011 meeting hosted by the Federal Reserve Bank of San Francisco discussed "Advancing Social Impact Investments through Measurement." It worked through many of the issues with the current measurement landscape. The main barriers to impact measurement highlighted were the lack of longitudinal data; benchmarks; third-party verification; transparent methodologies; and publicly available and voluntarily provided reporting [52].

Most of the systems discussed here focus on performance data from developing countries, and more work will need to be done to truly standardize the field. Yet, overcomplicating things is not the answer either. Large, institutional investors need clear, actionable data to join in on impact investments, at acceptable cost. Increased impact reporting through greater transparency is the key to scale up the impact investment market. However, perfection can be the enemy of the good. To help strike the right balance, governments and philanthropic funders can help respond to the limitations of current leading systems by incentivizing harmonization and emergence of measurement standards apt to support institutional investment. With greater disclosure of returns—both social and financial—of impact investments, a large-scale marketplace can begin to develop and flourish. Better infrastructure leveraging information technology for its systems and processes across investment platforms is needed to deliver a superior customer experience at low cost. This will be instrumental in driving a global impact investment marketplace. Creating standards and establishing norms helps set the stage for further development by actors at all levels of the impact investment landscape, including mainstream businesses. Traditional businesses and investors all have a role to play in supporting and driving these infrastructure developments, as they will eventually benefit financially from the increased market opportunity. Fortunately, progress in information technology is continuously reducing the cost of information gathering and management, so that digitization should ultimately lend a helping hand to the emergence of cost-effective impact measurement methodologies.

Today, impact investment as a field of practice stands at an inflection point. Either it will address the critical barriers still limiting broader growth and capitalize on its amazing potential—or, if history is any indication, it will settle on the margins of conventional investment and be replaced by a new term and concept sometime in the next five to ten years. The opportunity is massive. To act on it, it is now time to ask impact investing to move to its release 2.0 where it can play at a scale that has transformational impact on financial markets. Next to product innovation, this means engaging with the right investment content in scale, and lowering the cost of impact reporting through technology and intelligent standardization.

In *Finance and the Good Society*, Robert Shiller, one of the only experts to predict the financial crisis, makes the case that what is needed now is more financial innovation, not less [53]. He states, "The financial crisis was not due simply to greed or dishonesty of players in the world of finance; it was ultimately due to fundamental structural shortcomings in our financial institutions" [53]. The irony, as Shiller emphasizes, is that "better financial instruments, not less activity in finance, is what we need to reduce the probability of financial crises in the future" [53].

Impact investing brings the concept of our values as expressed by social and environmental outcomes targeted to a world of investing that is being reshaped by digital and regulation. Making impact investible is the biggest opportunity in finance in at least a generation. The adjustment needs in our sustainability century are real. If impact investment does its homework, and asserts itself as an investment style rather than an asset class, enabling investing 100 % of assets for impact across multiple asset classes, the ramifications for the overall competitiveness, prosperity and sustainability of the world's market economies could very well exceed the estimate of an asset class reaching 0.1 % of all financial assets by 2020 [54].

Scale has to be our goal if we aim to enlist private and public capital markets in funding the kinds of opportunities discussed in the preceding parts of the book. Supply chain upgrading and energy are multitrillion dollar industries and a placeholder for reconciling competitiveness with social and environmental performance at large. To upgrade capital markets, the most promising approach is considering the yin and yang of pairing "purist" impact investment with the lean application of the most heuristically powerful impact principles to mainstream capital markets and asset classes. This to leverage fresh opportunities resulting from the changing structure of the financial industry resulting from digitization, as well as developing valid investment content that reconciles profit with social impact.

Fortunately, talents who want to make a professional contribution—the members of "Generation Impact" I mentioned at the outset of the book—are increasingly taking on the challenge of cracking the nut of 'profits with impact:' with the advent of impact investing, a new way of thinking about the conscious pursuit of financial and extra-financial goals is gaining ground. This should help to resource the energy transition and supply chain upgrading. There are rainy clouds out there, but life outside the casino has just started to look a bit brighter. If you believe in magic, the time to do the rain dance is now. Otherwise, it is best to roll up your sleeves and help a better investment paradigm get off the ground sooner rather than later—whether as an investor or a professional. Ready for life outside the casino? If you are in a position of formal authority, the time to empower the talent that reports to you is now—and do task yourself with contributing as well. Just imagine how powerful it would be if major institutional investors included greater emerging market exposure and forward looking upgrading leveraging the Internet of Things in their proxy voting strategies. This with an "impact light" mindset that carefully avoids offloading enormous layers of reporting costs on companies and financial institutions. The time to bring the power of financial innovation to bear on the biggest challenges of our time is now.

References

1. BlackRock (2014) Dealing with divergence 2015 investment outlook. BlackRock Investment Institute. http://www.blackrock.com/investing/library. Accessed 15 May 2015
2. Meadows DH et al (1972) The limits to growth. Universe Books, New York
3. BlackRock (2015) Engagement, proxy voting and ESG investment integration. http://www.blackrock.com/corporate/en-us/about-us/responsible-investment/engagement-and-proxy-voting . Accessed 15 May 2015
4. General Board Pension and Health Benefits (2015) An introduction to proxy voting. http://www.gbophb.org/investments/resources/introduction-to-proxy-voting. Accessed 15 May 2015
5. United States Senate Accountability Office (2004) Pension plans additional transparency and other actions needed in connection with proxy voting. https://books.google.ch/books?id=3rtgke4_TmwC&pg=PP2&lpg=PP2&dq=pension+plan+fiduciary+proxy+voting+started+in&source=bl&ots=JwGccyzgEv&sig=2pzjVFpcG2aHp7ILkLI8tZSS7oQ&hl=en&sa=X&ved=0CB4Q6AEwAGoVChMIu7zxop-KxgIVydMUCh1apAA8#v=onepage&q=pension%20plan%20fiduciary%20proxy%20voting%20started%20in&f=false. Accessed 3 May 2015
6. U.S. Securities and Exchange Commission (2015) Final rule: delivery of proxy statements and information statements to households. https://www.sec.gov/rules/final/33-7912.htm. Accessed 15 May 2015
7. Global Reporting Initiative (2013) Making the connection: using the GRI G4 guidelines to communicate progress on the UN Global Compact principles. https://www.globalreporting.org/resourcelibrary/UNGC-G4-linkage-publication.pdf. Accessed 5 May 2015
8. McKinsey & Company (2010) Lions on the move: the progress and potential of African economies. http://www.mckinsey.com/insights/africa/lions_on_the_move. Accessed 3 May 2014
9. International Telecommunication Union-ICT (2014) ICT facts and figures—the world in 2014. http://www.itu.int/en/ITU-D/Statistics/Documents/facts/ICTFactsFigures2014-e.pdf. Accessed 27 May 2015
10. ONE (2015) Energy: quick facts. http://www.one.org/us/issues/energy. Accessed 27 May 2015
11. McKinsey & Company (2010) Lions on the move: the progress and potential of African economies. http://www.mckinsey.com/insights/africa/lions_on_the_move. Accessed 3 May 2014
12. Economist Intelligence Unit (2014) Ethiopia: poverty levels remain high. http://country.eiu.com/article.aspx?articleid=1591934543&Country=Ethiopia&topic=Economy&subtopic=_7. Accessed 4 May 2015
13. IMF (2014) The Federal Democratic Republic of Ethiopia: 2014 Article IV consultation—staff report; press release; and statement by the Executive Director for the Federal Democratic Republic of Ethiopia. https://www.imf.org/external/pubs/ft/scr/2014/cr14303.pdf. Accessed 5 May 2015

14. US State Government (2015) Ethiopia's Growth and Transformation Plan (GTP) at-a-glance. http://photos.state.gov/libraries/ethiopia/427391/PDF%20files/GTP%20At-A-Glance.pdf. Accessed 2 May 2015
15. Martin M (2014) Removing the bottlenecks on the road to middle income and making impact investible in: Myanmar Business Today. http://www.mmbiztoday.com/articles/removing-bottlenecks-road-middle-income-and-making-impact-investible. Accessed 3 May 2015
16. The Federal Democratic Republic of Ethiopia (2010) Growth and Transformation Plan (GTP) draft. http://www.ethiopians.com/Ethiopia_GTP_2015.pdf. Accessed 3 May 2014
17. CIA World Factbook (2015) Ethiopia. https://www.cia.gov/library/publications/the-world-factbook/geos/et.html. Accessed 3 June 2015
18. Dile, YT (2013) Hydrological response to climate change for Gilgel Abay River, in the Lake Tana Basin—Upper Blue Nile Basin of Ethiopia. http://www.plosone.org/article/info%3Adoi%2F10.1371%2Fjournal.pone.0079296. Accessed 3 May 2014
19. Vidal J (2003) Ethiopia's worst famine in 20 years. In: Guardian. http://www.theguardian.com/world/2003/apr/18/famine.ethiopia. Accessed 4 Mar 2014
20. Government of Ethiopia (2012) Assessing progress toward the Millennium Development Goals: Ethiopia MDGs report 2012. http://www.et.undp.org/content/dam/ethiopia/docs/Ethiopia%20MDG%20Report%202012_Final.pdf. Accessed 3 Mar 2014
21. World Bank (2013) Africa's pulse: an analysis of issues shaping Africa's economic future. http://www.worldbank.org/content/dam/Worldbank/document/Africa/Report/Africas-Pulse-brochure_Vol7.pdf. Accessed 3 May 2015
22. World Bank Group (2015) Myanmar: empowering people for inclusive growth. http://www.worldbank.org/content/dam/Worldbank/Publications/eap/myanmar-2015-2017-country-partnership-framework.pdf. Accessed 4 May 2015
23. Asian Development Bank (2014) Myanmar: unlocking the potential, country diagnostic study. http://adb.org/sites/default/files/pub/2014/myanmar-unlocking-potential.pdf. Accessed 7 May 2015
24. Accenture & The Rockefeller Foundation (2015) De-centralized electricity in Africa and Southeast Asia: issues and solutions. https://www.rockefellerfoundation.org/app/uploads/De-centralized-Electricity-in-Africa-and-Southeast-Asia.pdf. Accessed 3 May 2015
25. World Bank (2014) Land under cereal production. http://data.worldbank.org/indicator/AG.LND.CREL.HA/countries?display=graph. Accessed 3 May 2015
26. Kurzweil R (2010) How my predictions are faring. http://www.kurzweilai.net/images/How-My-Predictions-Are-Faring.pdf. Accessed 3 May 2014
27. Lahoz-Beltra R (2014) The 'crisis of noosphere' as a limiting factor to achieve the point of technological singularity. http://arxiv.org/abs/1405.3378v1. Accessed 5 May 2015
28. Technology Review (2013) The year of the Internet of Things. http://www.technologyreview.com/view/509546/2013-the-year-of-the-internet-of-things. Accessed 3 May 2014
29. Evans D (2011) The Internet of Things: how the next evolution of the Internet is changing everything. http://www.cisco.com/web/about/ac79/docs/innov/IoT_IBSG_0411FINAL.pdf. Accessed 3 May 2014
30. Barbier J et al (2014) Internet of everything in ASEAN. http://www.cisco.com/web/about/ac79/docs/IoE/IoE-in-ASEAN.pdf. Accessed 3 May 2014
31. McKinsey & Company (2014) Resource revolution: gathering force. http://www.mckinsey.com/insights/mckinsey_quarterly/digital_newsstand/2014_number_2. Accessed 4 May 2015
32. PWC (2013) Internet of Things: evolving transactions into relationship. http://www.pwc.com/us/en/technology-forecast/2013/issue1/assets/techforecast 2013-issue-1.pdf. Accessed 3 May 2014
33. Evans D (2011) The Internet of Things: how the next evolution of the Internet is changing everything. http://www.cisco.com/web/about/ac79/docs/innov/IoT_IBSG_0411FINAL.pdf. Accessed 3 May 2014

34. Bradley J et al (2013) Embracing the Internet of everything to capture your share of $14.4 trillion. http://www.cisco.com/web/about/ac79/docs/innov/IoE_Economy.pdf. Accessed 3 May 2014
35. PWC (2013) Internet of Things: evolving transactions into relationship. http://www.pwc.com/us/en/technology-forecast/2013/issue1/assets/techforecast-2013-issue-1.pdf. Accessed 3 May 2014
36. Perera C et al (2013) Context aware computing for the Internet of Things: a survey. IEEE Communication Surveys and Tutorials(10)10. http://arxiv.org/pdf/1305.0982.pdf. Accessed 3 May 2014
37. Leary J, Kaczmarska M (2014) IoT cloud connect: Cisco IoT platform for service providers. http://www.ciscoknowledgenetwork.com/files/475_10-28-2014-IoT_Cloud_Connect-CKN_October_28_2014_v2.pdf?PRIORITY_CODE. Accessed 3 May 2015
38. Barbier et al (2014) Internet of everything in ASEAN. http://www.cisco.com/web/about/ac79/docs/IoE/IoE-in-ASEAN.pdf. Accessed 3 May 2014
39. Mitchell S et al (2013) The Internet of everything for cities: connecting people, process, data and things to improve the livability of cities and communities. http://www.cisco.com/web/strategy/docs/gov/everything-for-cities.pdf. Accessed 3 Mar 2014
40. CISCO (2012) City transforms economic sustainability with public cloud. http://www.cisco.com/c/dam/en/us/solutions/collateral/collaboration-endpoints/cisco_busan.pdf. Accessed 3 Mar 2014
41. Bradley J et al (2013) Embracing the Internet of everything to capture your share of $14.4 trillion. http://www.cisco.com/web/about/ac79/docs/innov/IoE_Economy.pdf. Accessed 3 May 2014
42. Lahoz-Beltra R (2014) The 'crisis of noosphere' as a limiting factor to achieve the point of technological singularity. http://arxiv.org/abs/1405.3378v1. Accessed 5 May 2015
43. Shahan Z (2014) 13 Charts on solar panel cost and growth trends. http://cleantechnica.com/2014/09/04/solar-panel-cost-trends-10-charts. Accessed 4 May 2015
44. IEA (2014) Technology roadmap solar photovoltaic energy. https://www.iea.org/publications/freepublications/publication/pv_roadmap.pdf. Accessed 4 May 2015
45. PWC (2013) Internet of Things: evolving transactions into relationship. http://www.pwc.com/us/en/technology-forecast/2013/issue1/assets/techforecast-2013-issue-1.pdf. Accessed 3 May 2014
46. Rifkin J (2014) The rise of anti-capitalism. In: New York Times. http://www.nytimes.com/2014/03/16/opinion/sunday/the-rise-of-anti-capitalism.html?_r=0. Accessed 3 May 2014
47. Bradford DeLong J, Summers LH (2001) The new economy: background, questions, speculations. http://citeseerx.ist.psu.edu/viewdoc/download?doi=10.1.1.15.3529&rep=rep1&type=pdf. Accessed 3 Mar 2014
48. Rifkin J (2013) Nord-Pas de Calais: Third Industrial Revolution Master Plan 2013. TIR Consulting Group, LLC. http://en.calameo.com/books/0028209601062e1413c26. Accessed 3 May 2015
49. Simon M (2015) Managing vs. measuring impact investment. http://www.ssireview.org/blog/entry/managing_vs_measuring_impact_investment. Accessed 21 Jul 2015
50. Clark C et al (2012) The impact investors: the need for evidence and engagement. Pacific Community Ventures, Inc., Impact Assets, and Duke University's Fuqua School of Business. http://www.pacificcommunityventures.org/uploads/reports-and-publications/The_Impact_Investor_201203.pdf. Accessed 1 May 2013
51. Tuan, MT (2011) Impact capital measurement: approaches to measuring the social impact of program–related investments. Robert Wood Johnson Foundation. http://www.rwjf.org/content/dam/web-assets/2011/11/impact-capital-measurement. Accessed 15 May 2015

52. Hornley B et al (2011) Advancing social impact investments through measurement conference: summary and themes. In: Community Development Investment Review. http://www.frbsf.org/publications/community/review/vol7_issue2/Review_vol7_issue2.pdf. Accessed 1 May 2013
53. Shiller RJ (2012) Finance and the good society. Princeton University Press, Princeton
54. Saltuk, Y. and El Idrissi, A (2015) Global social finance. Spotlight on the market: the impact investor survey. http://impactinvestingaustralia.com/countriesandregions/uk/j-p-morgan-giin-5th-annual-impact-investor-survey-launched. Accessed 27 May 2015

Epilogue: Let's Do It!

When the story of these times gets written, we want it to say that we did all we could, and it was more than anyone could have imagined.

Bono [1]

Our Future, Yea or Nay

To grasp that rendering the global economy pretty much closed-loop and powered mainly by renewable energy is the best long-run bet to sustain the projected population of nine to eleven billion people throughout this millennium, mastery of rocket science is not required. To avoid an overshoot and collapse pattern, we need to act systemically and decisively [2].

In spite of what people discuss at the 2015 UN climate change conference in Paris, and elsewhere, we do not yet have the magic recipe though [3]. Some ingredients are still missing. Considering what is technically and organizationally possible if we dare and are diligent, this book has attempted to chart a path forward. The gigatons of CO_2 emissions continue to pile up. For fixing energy, there is only one logical conclusion: to truly compete, renewable energy will have to become fully price competitive and substitutable with fossil fuels much faster.

My hope is that the preceding analysis brought home the point that real solutions are not mere science fiction. Just think about the energy cost and availability implications of high-efficiency solar energy harvesting leveraging a wide spectrum of light during the day and possibly even conducting infrared harvesting at night. Diligently applying science, there are many such possibilities that may seem out of the box only at first sight. For example, what if wind turbines were routinely coated in ways that would enable them to much better deal with turbulence so as to produce energy at lower wind speeds and suffer from much less downtime [4]? Again, this is in principle possible, merging design thinking from bionics with advances in materials sciences. After all, birds can do this—and so can we with a little effort.

M. Martin, *Building the Impact Economy*,
DOI 10.1007/978-3-319-25604-7

The preceding chapters pointed to one key challenge that stands in the way of devising a viable energy path forward: the innovation cycle is still too long. To stay with solar photovoltaics for a minute, it took more than 100 years from Becquerel's discovery of the photovoltaic effect in 1839 to Bell Lab unveiling the first usable silicon solar cell in 1954, with a 6 % efficiency [3].

Granted, we have gotten better since. We also do not have infinite amounts of time to get it right. Existing innovation programs are often not sufficiently agile. In energy, it is now time to make the energy transition real. This can only happen if breakthroughs in science reach full-scale impact much faster. A 100 % renewable energy economy in our lifetime can come into view—if we adjust our speed.

A Closer Look at Human Nature

To make that happen, it is key that we understand human nature and use that knowledge to advantage. Throughout human history, managing problems and finding solutions has been the normal state of affairs. Our ability to do so affects the course of history. When the Plague of Justinian the Great wiped out perhaps as many as 25 million people in the East Roman Empire in 541–542—just when Justinian's armies had retaken most of Italy and the western Mediterranean coast—, it helped to finish off the Mediterranean world of late Antiquity as we know it. Today, global megatrends rearrange reality with a truly amazing force. They affect formal contexts, individual identities, and group dynamics on a scale that seems historically unprecedented—and, given the environmental dimension, in ways that could be irreversible.

As global change raises interdependence, the complexity of the global system increases. In itself, complexity is nothing new. Complexity is both a result and a driver of the division of labor. The division of labor provides the basis for contemporary prosperity. Expert systems or "systems of technical accomplishment or professional expertise that organize large areas of the material and social environments in which we live today," are fundamental to human progress [5]. Many new expert systems have emerged to handle this complexity since the advent of industrialization, and new ones are forming. "Sustainability" and "sustainable finance" or "impact investing" are examples of such newer emerging expert systems. Many established models inspire less confidence, opening up space for fresh approaches.

Technological and scientific progress puts potent new instruments at people's disposition. Whether we are talking about blockchains in finance, swarms of batteries stabilizing a renewables-powered grid, or chemical change management in apparel, in all industries discussed in this book, the power of digital paired with hardware technology holds the prospect of much higher performance at lower cost, driving both profits and social impact—if we get it right.

We witness an explosion of knowledge. Technology that helps to gather, process, and disseminate information is increasingly sophisticated and available, all at lower cost. Of course, technological and scientific progress comes with a cost attached—it also manufactures new risks such as computer viruses or biological weapons. The possibility to simply upscale existing expert systems to manage ever-higher levels of complexity is also about to reach some limit.

As we go about devising the new solutions needed, our best hedge against both excessive optimism and undue pessimism is a closer look at human nature. If we keep it simple and see humans as governed by rational choice, we foreground that we need to get governance structures and incentive systems right [6]. This is certainly true, but at best a partial answer. Alternatively, we could zoom in on conflict. Strife is the father of all things. Open conflict often unleashes energy for catalytic change. Covert conflict is more widespread than is apparent at first sight. We would not have the Internet without US Department of Defense sponsored research under the DARPA program [7].

In spite of the potency of the rational choice and conflict perspectives, we must not forget that the world is not merely a cold place in which everything revolves around the control and allocation of resources. Deep down, people are often non-rational and driven by their subconscious. We have a need for the sacred. Symbols and rituals play an important role [8]. Moreover, expert statements exert an influence on defining what we perceive to be reality, and which forces we believe to drive the transformation of reality.

Don't Take Representations of Reality as a Given

Many environmentalists argue that humanity makes the planet ill. Some argue humanity has gone hysteric with all the new products and digitization. Well, how would a medical doctor have cured a hysteric 250 years ago? Take Dr. Pomme's therapy, narrated by French social theorist Michel Foucault [9, 10]. Pomme was a mid-eighteenth-century doctor who treated and cured a hysteric. The success of the therapy seems surprising by today's standards, given the treatment: Pomme made his patient take "baths, ten or twelve hours a day for ten whole months." Those who have a graphical mind may imagine what this does to someone's skin. The objective of the treatment was the dessication of the nervous system and the heat that sustained it. At the end of the treatment, Pomme saw "membranous tissues like pieces of damp parchment [...] peel away with some slight discomfort" [10].

I will stop here to spare you the therapy's less pleasant aspects. Pomme was not a recent graduate of Harvard Medical School. All of *us* would interpret his medical cure as an elaborate ritual of torture. The point though is that the treatment made perfect sense in the eighteenth century. At the time, medical discourse provided Pomme's contemporaries with a set of categories and interpretations that made it seem perfectly "normal" to have someone bathe ten hours a day for ten months in a row to cure hysteria. By defining appropriate objects and practices of intervention, medical discourse contributed to constructing a reality that made sense to people.

What does this mean for us today? We have our job cut out: to make the sustainable economy the new normal. This in ways that also create prosperity and provide space for private initiative. For Foucault, discourses completely determine what people think, know and do. This seems exaggerated. The more general point is that discourses play a paramount role in shaping reality. This is why we scouted the solutions for the 80–100 % renewable economy [4]. If we do not set such a goal and construct a technology scouting logic accordingly, we will not find and nurture these technologies. They risk remaining invisible to us.

Discourses are not merely interesting as a historical phenomenon. They play a role in shaping contemporary action that we must look at. Today, expert statements are more important and pervasive than previously.

Commit to Progress at the Bleeding Edge

We need to commit to achieving progress at the bleeding edge, and the way we organize business and the economy has profound consequences. The key message of this book is that we can succeed at scale—provided we structure these efforts as a winning formula for business as well. For sure, today's matters are more complex than Dr. Pomme's approach would suggest. We can choose between different types of medicine, but so far they have turned out to be painkillers only.

When we think about human nature and its consequences for human action, let's not forget that the world is not just composed of hard-edged, objective structures, cognitive or otherwise, as many observers of society have pointed out. It is useful to have the symbolic interactionist school's three "classical" assumptions on the radar [11, p. 2]: First, people act toward things on the basis of the meanings that the things have for them. Second, meaning arises out of social interaction. Finally, as people encounter things, they make and handle meanings through an interpretive process—mindsets matter and decide what we make of new possibilities. For example, as robots replace people with an adverse impact on the tax base, should businesses pay payroll taxes and social security on robot workers [12]? As scientific luminary Stephen Hawking warns about the potential dangers of artificial intelligence, will we soon need a Geneva Convention type arrangement that bans artificial intelligence in military systems [13]?

The power of mindsets is an insight our work has corroborated, for example in upgrading manufacturing practices in the textile and garment industry: the most difficult part is not the discovery of hidden efficiency potentials that enable higher economic, environmental and social performance [14]. Rather, the most difficult part is to inspire a sense of daring so we do take on the challenge, and to shift mindsets so people can become emotionally (and resource-wise) invested in what can actually become superior practice.

When Galileo started promoting the notion that the Earth and planets revolved around the sun at the center of the solar system in 1610, the heliocentric worldview was almost 2000 years old. It would take another 382 years until Pope John Paul II validated his insight, when he argued:

> Thanks to his intuition as a brilliant physicist and by relying on different arguments, Galileo, who practically invented the experimental method, understood why only the sun could function as the center of the world, as it was then known, that is to say, as a planetary system. The error of the theologians of the time, when they maintained the centrality of the Earth, was to think that our understanding of the physical world's structure was, in some way, imposed by the literal sense of Sacred Scripture [15].

The analysis of the role impact investing could play in modernizing capital markets has similarly shown that bold new concepts are instrumental in taking better practice forward. They could give a major push, including to the energy transition and the lives of people in emerging markets.

Today, we need to move things faster and attract a growing audience, leading to superior outcomes in practice. Corporate innovation, long supply chains, energy and capital markets are all theaters in which we could converge on a modus operandi that systematically produces such improved outcomes. Entertaining a series of *how-to* questions, the preceding chapters have shown concrete ways how to go about this.

At this stage, it is certainly not a surprise for you that my personal vote on our future is "yea." I hope you are by now equally more firmly convinced that the "our future, yea" camp is actually viable—and that you will join along. Perhaps you are already actively assessing how to apply the many practical recommendations and examples in this book to your own practice.

When our descendants look back at the first quarter of the twenty-first century in 2115, let's make sure these improvements are so dramatic that they will have mattered on the path to building the Impact Economy. We should never underestimate the power of our ideas and our ability to create the preferred future we would like to inhabit.

Let's do it!

References

1. Bono (2004) US senate hearings and testimony: foreign operations subcommittee hearing on the FY05 budget request for HIV/AIDS. http://www.atu2.com/news/us-senate-hearings-and-testimony-testimony-of-bono-founding-member-data.html. Accessed 15 May 2015
2. Overpopulation.org (2010) The overpopulation register. http://www.overpopulation.org/consumption.html. Accessed 15 May 2015
3. Martin M (2014) The time to shorten the innovation cycle and make the energy transition real is now. In: CleantechIQ. http://cleantechiq.com/2014/12/the-time-to-shorten-the-innovation-cycle-and-make-the-energy-transition-real-is-now/. Accessed 30 May 2015
4. Martin M (2015) Energy transition fast forward! Scouting the solutions for the 80–100% renewable economy. Impact Economy. http://www.impacteconomy.com/papers/IE_PRIMER_APRIL2015_EN.pdf. Accessed 15 May 2015
5. Giddens A (1990) The consequences of modernity. Stanford University Press, Stanford

6. Ederer F, Manso G (2011) Incentives for innovation: bankruptcy, corporate governance and compensation systems. In: Litan RE (ed) Handbook on law and growth, Edward Elgar, Cheltenham. http://www.elgaronline.com/view/9781848441873.00009.xml. Accessed 15 July 2015
7. Leiner B et al (2015) Brief history of the internet. http://www.internetsociety.org/internet/what-internet/history-internet/brief-history-internet. Accessed 15 May 2015
8. Carter J (2009) The role of the arts in worship. http://www.rpinet.com/ml/2805arts.html. Accessed 30 May 2015
9. Foucault M (1975) The birth of the clinic: An archaeology of medical perception. Harper Colophon, New York
10. Dreyfus HL, Rabinow P (1983) Michel Foucault: Beyond structuralism and hermeneutics, 2nd edn. University of Chicago Press, Chicago
11. Blumer H 1986 (1969) Symbolic interactionism: perspective and method. University of California Press, Berkeley and Los Angeles, p 2
12. Yakabuski K (2014) Taxes pay for robots, but robots don't pay taxes. In: The Globe and Mail. http://www.theglobeandmail.com/globe-debate/taxes-pay-for-robots-but-robots-dont-pay-taxes/article17942327. Accessed 23 July 2015
13. Foreman A (2015) Ancient automatons, modern worries. In: Wall Street Journal. http://www.wsj.com/articles/ancient-automatons-modern-worries-1425577257. Accessed 23 July 2015
14. Martin M (2013) Creating sustainable apparel value chains. Impact Economy. http://www.impacteconomy.com/papers/IE_PRIMER_DECEMBER2013_EN.pdf. Accessed 15 May 2015
15. Robinson DN (2006) Science and faith: the warrants for belief. In: Robinson, DN et al (eds) Human nature in its wholeness: a roman catholic perspective. Catholic University of America Press, Washington, DC: 169. https://books.google.ch/books?id=jB3ePe8DJuAC&pg=169&redir_esc=y#v=onepage&q&f=false. Accessed 19 July 2015

MIX
Papier aus verantwortungsvollen Quellen
Paper from responsible sources
FSC® C105338

If you have any concerns about our products,
you can contact us on
ProductSafety@springernature.com

In case Publisher is established outside the EU,
the EU authorized representative is:
Springer Nature Customer Service Center GmbH
Europaplatz 3, 69115 Heidelberg, Germany

Printed by Libri Plureos GmbH
in Hamburg, Germany